YOUR GUIDE TO PROPAGATION

Ian Poole, G3YWX

Radio Society of Great Britain

Published by the Radio Society of Great Britain, Cranborne Road, Potters Bar, Herts EN6 3JE.

First published 1998

© Radio Society of Great Britain, 1998. All rights reserved. No part of this publication may be reproduced, stored in a retrieval system, or transmitted, in any form or by any means, electronic, mechanical, photocopying, recording or otherwise, without the prior written permission of the Radio Society of Great Britain.

ISBN 1 872309 49 6

Publisher's note

The opinions expressed in this book are those of the author and not necessarily those of the RSGB. While the information presented is believed to be correct, the author, the publisher and their agents cannot accept responsibility for consequences arising from any inaccuracies or omissions.

Cover design: Jennifer Crocker.
Illustrations: Roy Pettit and Bob Ryan.
Production: Mike Dennison.
Typography: Ray Eckersley, Seven Stars Publishing.

Printed in Great Britain The Nuffield Press Ltd, Abingdon.

Contents

Preface iv

1 Introduction to radio wave propagation 1

2 Radio waves 6

3 The atmosphere and the Sun 14

4 Ionospheric propagation 27

5 Ionospheric storms and auroras 39

6 Predicting, assessing and using HF propagation 47

7 Other ionospheric modes 57

8 Tropospheric propagation.................. 65

9 Meteor scatter 72

Appendix: Further reading and sources of data... 80

Index..................................... 81

PREFACE

For radio amateurs, short wave listeners and anyone involved in radio, a knowledge of radio wave propagation is almost essential. It enables them to be able to use their equipment to the best, and it is particularly useful for DXing, whether on the HF, VHF or UHF bands. By knowing how to utilise the way the signals propagate, the best times and methods can be adopted, making the possibility of success much greater.

The subject of radio propagation is fascinating, and I hope that you enjoy reading the book as much as I have enjoyed researching for it and writing it. However, the book would not have been possible without encouragement from others. I would particularly like to thank Marcia Brimson and Mike Dennison at the RSGB, and also Ray Eckersley, for their help.

Ian Poole
October 1998

Chapter 1
INTRODUCTION TO RADIO WAVE PROPAGATION

Radio wave propagation is at the very heart of radio technology. By understanding the way that radio waves travel it is possible to make the best use of them.

On some frequencies signals may travel little further than the line of sight, and on other frequencies the distances can extend to just over the horizon, whilst on others communications are possible over the whole of the globe. Each type of propagation has its advantages. Some are better for local communications because there is less interference from distant stations whereas others give the possibility of communicating with people in a wide variety of interesting places.

Whatever one's interest in radio, the propagation of signals is of great importance. For the keen DXer it is essential to have a good idea what each band offers and which one will give the best chance of making a contact or hearing stations from a particular area.

It is also very important to be able to have a feel for the conditions at any given time. By knowing how the bands perform it is possible to use them to their best. Often a person who knows the conditions well will have the best chance of being able to hear or make contact with the most interesting stations.

In order to know how the different bands are likely to perform at different times it is necessary to have a basic knowledge about propagation. Different frequencies are influenced by different effects. In turn these effects vary with time, which can result from changes in the weather, right up to changes in the time of day, season and even the state of the Sun itself. The experienced DXer will be able to take these into account and judge when and on which frequency to listen.

The study of propagation is a fascinating topic for any radio enthusiast. Not only can it pay dividends in terms of operating, but also it can become an absorbing additional interest in the hobby.

RESEARCH WORK

It has taken many years to gain a reasonable understanding of the reasons why signals propagate in the way they do. Many of the effects are the result of complicated chains of events. As a result of this, even today's science does not have a complete understanding of every aspect of radio wave propagation. However, the knowledge we do have has been the result of the work and observations of many people. Often radio amateurs have been at the forefront of the research, and in the very early days of radio they helped discover the importance of the short-wave bands, making the first contacts across the Atlantic on wavelengths below 200m.

When radio was first discovered, people had little idea of how far the signals could travel. Amazement was expressed when it was shown that radio waves could pass through walls. When Marconi started his investigations into radio he steadily increased the distance over which he could make communication. He also investigated many aspects about radio signals which we take for granted today, including whether they could travel over water.

Initially it was thought that radio signals would not be able to travel much further than the line of sight. However, in 1899 when Marconi made a transmission across the channel from Dover to Boulogne his engineers were very surprised to hear the transmissions back at his factory in Chelmsford. This was a distance of over 130km and well beyond the line of sight. This prompted Marconi to consider the possibility of communications across the Atlantic. In 1901 after many setbacks he managed to receive a signal and made the first transmission over such a large distance.

Around this time two scientists, Kennelly in the USA and Heaviside in the UK, independently put forward the idea that there were conducting layers in the upper atmosphere that could guide radio waves over considerable distances. As a result of this research the layers in the ionosphere were termed the *Kennelly-Heaviside*, or *Heaviside* layers.

Further work was undertaken by a number of people in the years between 1910 and 1920. An American named de Forest, one of Marconi's main competitors in the commercial communications field, suggested that the reflecting layers in the upper atmosphere were at a height of about 100km. However, it took until 1924 before the British scientist Appleton confirmed the existence of the layer. He named the area the *'E' layer*, and a year later discovered another layer above it which he denoted by the letter 'F'.

Despite these early advances comparatively little was understood about the way in which signals travelled. It was thought that it was only possible for long-distance transmissions to be made on long wavelengths. Radio amateurs were largely responsible for discovering the value of the short-wave bands for long-distance communications. After the First World War they were relegated to wavelengths shorter than 200m but many American amateurs found they could still make contacts over considerable distances. It took a number of attempts before two-way transatlantic contacts were made, but the first contact took place in November 1923 between 1MO and 1XAM in the USA and (F) 8AB in France. Naturally further contacts soon followed and it did not take long before greater distances were reached. In October 1924, G2SZ, the station of the Mill Hill School in North London, managed to make contact with ZL4AA in New Zealand, proving that worldwide communication was possible on the short-wave bands.

At the same time professional organisations were beginning to explore these wavelengths and soon they also discovered that they supported long-distance communications. In one experiment Marconi used his yacht Elettra to receive signals and plot the strengths as it sailed away from the transmitting station in Britain. Others as well started to exploit these frequencies and soon long-distance links began to be set up.

Scientists also started to gain a greater understanding of the ionosphere around this time. In 1926 two researchers named Taylor and Hulbert described the changing properties of the reflecting layers. They noticed that they changed over the period of a day as well as with the seasons. The general way in which HF signals were reflected slowly started to be understood, and phenomena including the skip distance and multiple hops were recognised.

Another milestone was passed in 1931 when Chapman published a paper on the Kennelly-Heaviside layer. This contained the fundamental basis for much of today's understanding of the ionosphere. Other work was undertaken by others including

Appleton. He performed more theoretical work on discovering how radio waves behaved in areas such as the ionosphere. In the work he published in 1932 he used the term *ionosphere* to describe areas of high ion concentration, and in doing so he gave it the name we use today.

The way in which the ionosphere varied with the sunspot cycle was also investigated. Sunspots had first been seen by the ancient Chinese many years before the birth of Christ. The first records of sunspots were made after the invention of the telescope by Galileo in 1611. However, it was not until the middle of the 18th century that records of sunspot numbers were kept on a regular basis, and then in 1843 a scientist named Schwabe published his findings about the cyclical nature of the sunspot numbers. These discoveries were taken further by Edison Petit at the Mount Wilson Observatory in the USA. In the late 'twenties and early 'thirties he established a link between the sunspot number and the level of ultra-violet radiation being received from the Sun. This was a vital link in the understanding of the variations in the ionosphere.

Despite all these discoveries there were still many unknown factors about the ionosphere. In the late 'fifties and early 'sixties space exploration started. The probes allowed far more data to be collected about the outer reaches of the atmosphere. Up until this time detailed studies of the ionosphere were not possible. These areas were above the highest altitudes that could be reached by balloons or aircraft and little was understood about their composition and even less about ion formation. As the probes penetrated further into space this increased scientific understanding about these areas. It also increased an understanding of the areas beyond the ionosphere. Whilst they do not affect radio waves themselves, they affect the ionosphere and hence have a bearing on propagation conditions.

The study of radio propagation was not confined to the shortwave bands. From well before the Second World War, people were exploiting the frequencies above 30MHz, and along with this it has been necessary to discover how signals travel at these frequencies. As they are affected by different areas of the atmosphere it has lead to new areas of research, and new discoveries about signal propagation at higher frequencies.

INTRODUCTION TO RADIO WAVE PROPAGATION

TODAY

Today radio amateurs are still able to make significant contributions to the study of propagation. They have the advantage that at certain times large numbers of stations may be active and can make observations. This is particularly true when there are contests on the band. At times like these it is possible to put the results of the contest together and look at the paths over which transmissions have been made, and then analyse the results. This has been important in looking at the results obtained at frequencies of 30MHz and above.

Radio amateurs can also make a valuable contribution to the general understanding of propagation because they are often interested in modes that may not be of great interest to professional organisations. This research all adds to the overall understanding of radio wave propagation and it is of vital importance. One area where this is true is in the evaluation of a mode called *transequatorial propagation*. Without the vigilance and expertise of radio amateurs this would probably not have been seen, and certainly not monitored.

Although not everyone will want to take part in these investigations, the topic of radio propagation is a fascinating study. It will help in the general understanding of radio, and will enable any operator to make the most of the equipment he or she is using. Apart from this it an interesting topic to delve into, so beware!

Chapter 2
RADIO WAVES

Before looking at radio signal propagation around the Earth it is worth investigating what radio signals themselves are and how they travel in free space.

NATURE OF RADIO SIGNALS

When a signal is transmitted from an antenna it can be regarded as a succession of concentric spheres moving outwards with ever-increasing radius. A useful two-dimensional comparison is that of a pond into which a stone is dropped. Waves move outwards from the centre, gradually reducing in amplitude as they move away from it. In reality the transmitting antenna does not radiate equally in all directions. However, the amplitude of the waveform still reduces the further it moves away from the source.

An *electromagnetic field* is a complex waveform consisting of electric and magnetic fields that are inseparable from one another. The two components are always in phase and the ratio of their amplitudes remains constant. The electric field arises from changes in voltage, whilst the magnetic component is due to changes in current. The two fields are at right-angles to one another as shown in Fig 2.1.

Another important aspect of radio waves is their *wavelength*. This is the distance between a point on one cycle of a wave to the equivalent point on the next as shown in Fig 2.2. A convenient point to take for this description is the peak as it is easy to show.

In previous years the position of a station on a radio dial was given in terms of its wavelength. For example, the BBC transmission from Droitwich on the long wave was given as 1500m. Nowadays the position of a station is noted in terms of its *frequency*, ie the number of vibrations in a given time. The unit of frequency is the *hertz* (Hz) which corresponds to one cycle or vibration per second. As the frequencies which are used for radio transmissions are normally very high, the standard prefixes are used to give *kilohertz* (1000Hz or 1kHz), *megahertz* (1,000,000Hz or 1MHz) and *gigahertz* (1,000,000,000Hz or 1GHz).

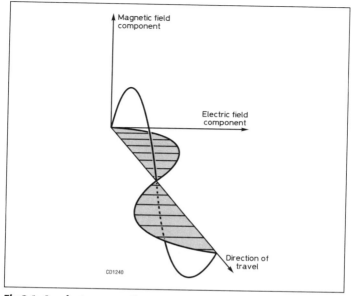

Fig 2.1. An electromagnetic wave

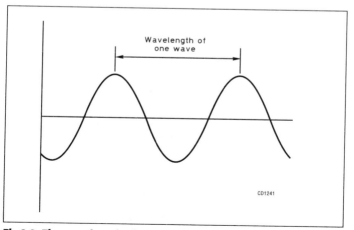

Fig 2.2. The wavelength of a wave

The *speed* of an electromagnetic wave is also important. This is normally taken to be 3×10^8m per second, although its exact value is 299,792,500m per second in a vacuum. If the wave travels in any other medium its velocity will be altered slightly.

The speed, frequency and wavelength of a wave are all related to one another. Accordingly it is possible to calculate the frequency if the wavelength is known and the wavelength if the frequency is known. The formula is:

$$v = \lambda \times f$$

where v is the speed of an electromagnetic wave (3×10^8m/s), λ is the wavelength in metres and f is the frequency in hertz.

Taking the example of a wavelength of 1500m, if this is substituted into the formula along with the speed of the electromagnetic wave it becomes:

$$3 \times 10^8 = 1500 \times f$$

Rearranging this, it gives:

$$f = \frac{3 \times 10^8}{1500} = 200,000 \text{Hz} \quad (200\text{kHz})$$

SPECTRUM

Electromagnetic waves cover a very wide spectrum, from the very-low-frequency radio transmissions at a few kilohertz right up to light waves, ultra-violet and beyond as shown in Fig 2.3. Within the radio portion of the spectrum there is still a vast range of frequencies. These too are split up into their various bands. Within the various sections of the radio spectrum are all the bands with which radio enthusiasts are familiar. The LF portion of the spectrum contains the long-wave broadcast band. The MF band contains the medium-wave broadcast band as well as the amateur radio 'Top Band'. Moving up in frequency, the HF section, extending from 3 to 30MHz, contains most of what are the traditional short-wave bands which include a variety of allocations such as amateur, broadcast, ship-to-shore and standard frequency transmissions. Above this is the VHF section. Here a variety of transmissions are found, ranging from VHF FM broadcasts to private mobile radio (PMR) systems as well as the popular 2m amateur band. Higher in frequency is the UHF band. In this

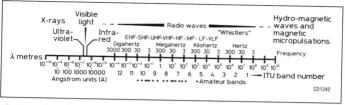

Fig 2.3. The electromagnetic spectrum

region are found the terrestrial television transmissions as well as cellular phone allocations and the 70cm amateur band. Above the UHF bands are found the SHF bands. Within this portion of the spectrum transmissions include those for direct broadcast satellite television.

From this it can be seen that the way in which radio signals travel around the Earth changes with frequency. Different techniques are used at the lower frequencies compared with those that are employed at higher ones. For example, most medium-wave broadcast stations are used for regional or local broadcasting over distances up to a few hundred kilometres, whereas short-wave stations are used for broadcasting over much greater distances.

POLARISATION

All electromagnetic waves can be said to have a certain *polarisation*. The polarisation of a wave is basically the plane in which the vibrations occur. Light waves show that this can be important in some circumstances. For example, sunglasses with Polaroid™ lenses only allow light of a particular polarisation through, and this can be used to reduce glare from one polarisation which arises from reflections.

Polarisation is also important for radio waves because antennas are generally only sensitive to signals with a given polarisation. In fact vertical antennas are sensitive to vertically polarised waves and similarly horizontal antennas are sensitive to horizontally polarised waves. They also radiate signals with the same polarisation as the antenna.

As electromagnetic waves have components that are at right-angles to one another, the polarisation is taken to be the polarisation of the electric component.

FIELD AND SIGNAL STRENGTH

Measurements or calculations of the strength of an electromagnetic wave sometimes need to be made. The intensity of a radio wave at any point in space may be expressed in terms of the strength of its electric field. This is expressed in terms of the electric force between points spaced one metre apart, ie in volts per metre.

It is found that as an electromagnetic wave travels away from the antenna it spreads out over an increasing area. Under free-space conditions (ie away from the influence of the Earth) it is easy to calculate what the signal strength will be because the signal follows an inverse square-law relationship. In other words, if the distance doubles the signal decays by a factor of two squared, ie four.

The graph shown in Fig 2.4 shows the calculated field strength levels when a transmitter radiates a signal into free space. If the antenna has a gain then this can be easily transferred onto the graph to give the correct levels.

Signals encountered by radio amateurs and listeners will not completely follow the free-space calculations because of the Earth itself as well as other objects in or close to the signal path. Conditions in the atmosphere, in particular the troposphere and the ionosphere, also have a significant effect. This makes it more difficult to predict the signal strength at a given point.

Calculating the field strength of a signal is normally of little use on its own. It is the actual signal voltage at the input of the receiver that is needed. Assuming that the antenna for the receiving system is a half-wave dipole and it is correctly matched to the receiver input, then the voltage input can be calculated from:

$$V = \frac{47.8 \times E}{f}$$

where V is the voltage at the input of the receiver in microvolts, E is the field strength at the antenna in microvolts per metre, and f is the frequency of operation.

If another type of antenna is used, then the gain or loss of the antenna can be used to modify the answer. In other words, if the antenna has a gain of 6dB over a dipole this represents a voltage increase by a factor of two, and the answer can be adjusted

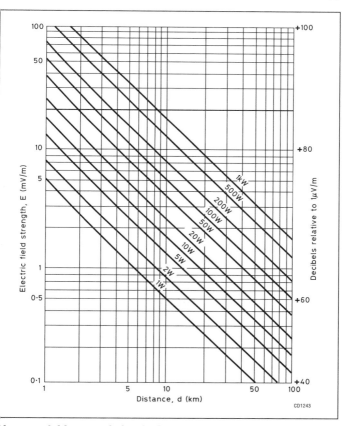

Fig 2.4. Field strength levels from an omni-directional antenna radiating into free space

accordingly. With knowledge of the performance of the radio receiver it is possible to determine whether the signal will be sufficiently strong to copy.

PATH LOSS

In many calculations it is necessary to determine the loss in a given path. Whilst it is not always easy to accurately determine the loss via an ionospheric path because of the great number of variables, paths in free space can be calculated quite accurately. It is quite common to calculate the path loss to and from a satellite. With the enormous distances to satellites in geostationary orbit (at altitudes of 35,860km above the Equator) losses caused purely by the distances involved are very large. By having a knowledge of the path loss, it is possible to optimise the requirements for receiver sensitivity, antenna gains and transmitter power. These are particularly important because increasing the transmitter power or the antenna size at the satellite can be very costly.

The loss can be calculated from the formula:

$$\text{Loss (dB)} = 32.45 + 20 \log_{10} f + 20 \log_{10} d$$

where f is the frequency in megahertz and d is the path distance in kilometres.

Notice that the loss increases with frequency. Fortunately antenna gains can be increased more easily at higher frequencies as antenna sizes are smaller, and this can be used to compensate this effect.

This means that if someone was directly below a satellite in geostationary orbit then the signal loss to the satellite at a frequency of 10GHz would be:

$$\begin{aligned}\text{Loss} &= 32.45 + 20 \log_{10}(10000) + 20 \log_{10}(35860) \\ &= 32.45 + 20 \times 4 + 20 \times 4.55 \\ &= 203.45 \text{dB}\end{aligned}$$

In view of this enormous loss geostationary satellite systems need to be carefully designed to ensure that a sufficiently good signal can be received. Systems using satellites in a lower orbit are not nearly as critical because path losses are very much lower.

MODES OF PROPAGATION

Although many of the properties of radio signals which have been mentioned in this chapter relate to the idealised situation when the signal is not modified by external influences, this is not the case for most circumstances. The signals are affected in a number of ways, and this governs the way they travel. At some frequencies the signals may only travel over short distances whereas at others they may be heard over distances of many thousands of kilometres. The way in which signals propagate can be split into four main categories as detailed below.

The most obvious form of propagation is *free-space* propagation. Here the radio signals travel in free space, ie away from other objects which influence the way in which they travel. It is only the distance from the source that affects the way in which the field strength reduces. This type of propagation is encountered with signals travelling to and from satellites.

The second type is where signals travel via *ground-wave* propagation. When signals travel via the ground wave they are modified by the ground or terrain over which they travel. They also tend to follow the Earth's curvature. Signals heard on the medium-wave band during the day use this form of propagation.

The third type is called *ionospheric* propagation. Here the radio signals are modified and influenced by the action of the free electrons in the upper reaches of the Earth's atmosphere called the *ionosphere*. This form of propagation is used by stations on the short-wave bands to enable their signals to be heard around the globe.

Finally there is *tropospheric* propagation. Here the signals are influenced by the variations of refractive index in the *troposphere* just above the Earth's surface. Tropospheric propagation is often the means by which signals at VHF and above are heard over extended distances.

Chapter 3
THE ATMOSPHERE AND THE SUN

Radio waves can travel over vast distances in outer space if they follow a line-of-sight path. However, they are able to travel around the Earth as a result of a variety of effects which take place in areas of the atmosphere. To understand how this happens it is necessary to take a look at the different areas of the atmosphere and how they can affect radio waves.

LAYERS OF THE ATMOSPHERE

The atmosphere can be split up into a variety of different layers according to their properties. However, as different areas of science look at different properties there is no single nomenclature for the layers. The system which is most widely used is that associated with meteorology and this is shown in Fig 3.1. From this it can be seen that the *troposphere* is the part of the atmosphere closest to the ground, extending to a height of 10km. The *stratosphere* is found above this at altitudes between 10 and 50km. This contains the ozone layer at a height of around 20km. Above the stratosphere, there is the *mesosphere* extending from an altitude of 50km to 80km, and above this is the *thermosphere* where temperatures can reach anything up to 1200°C.

The two main layers that are of interest from a radio viewpoint are the troposphere, which tends to affect frequencies above 30MHz, and the ionosphere. This is a region that crosses over the boundaries of the meteorological layers and extends from around 60km up to 700km. Here the air becomes ionised, producing ions and free electrons. The free electrons affect radio waves, often bending them back to the ground so that they can be heard over vast distances around the world.

Troposphere

This is the lowest of the layers of the atmosphere, and it is within this region that the effects that govern our weather occur. Low clouds occur at altitudes of up to 2km whereas medium level clouds extend to about 4km. The highest clouds are found a

THE ATMOSPHERE AND THE SUN

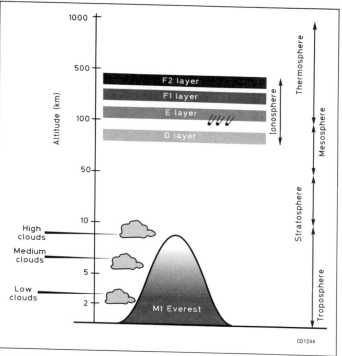

Fig 3.1. Layers of the atmosphere

altitudes up to 10km whereas modern jet airliners fly above this at altitudes of up to 15km.

Within the troposphere there is a steady fall in temperature with height and this has a distinct bearing on some propagation modes which occur in this region. This fall continues in the troposphere until the *tropopause* is reached. This is the area where the gradient levels out and then the temperature starts to rise. At this point the temperature is around −50°C.

The refractive index of the air in the troposphere plays a dominant role in radio signal propagation. This depends upon the temperature, pressure and humidity. When radio signals are affected this often occurs at altitudes up to 2km.

The ionosphere

In most regions of the atmosphere the molecules are in a combined state and remain electrically neutral. However, the ionosphere consists of an area where there is a large concentration of free positive ions and negative electrons. Of these it is the free electrons which actually affect radio waves. These electrons start to be found at an altitude of around 30km. However, it is not until an altitude of approximately 60km is reached that the number of them rises to sufficient levels to have a significant effect on radio signals.

The formation of the ionosphere is a very complicated process. It involves radiation from the Sun striking the molecules in the upper atmosphere. This radiation is so intense that when it strikes the gas molecules some electrons are given sufficient energy to escape the molecular structure (Fig 3.2). This leaves a molecule with a deficit of one electron, which is called an *ion*, and a free electron. As might be expected, the most common molecules to be

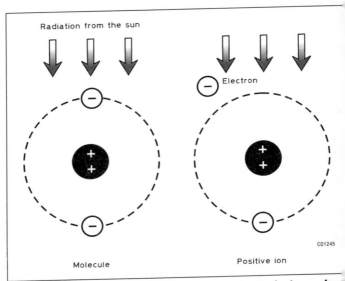

Fig 3.2. Radiation from the Sun strikes molecules in the ionosphere to form ions and electrons

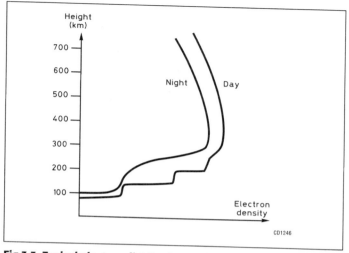

Fig 3.3. Typical electron distribution at day and night

ionised are nitrogen and oxygen, which even in this region of the atmosphere still have the highest concentration.

It is generally the ultra-violet light which causes this *ionisation* to occur. At very high altitudes the gases are very thin and only low levels of ionisation are created. As the radiation penetrates further into the atmosphere the density of the gases increases and accordingly the numbers of molecules being ionised increase. However, when molecules are ionised the energy in the radiation is reduced, and even though the gas density is higher at lower altitudes the degree of ionisation becomes less because of the reduction of the level of ultra-violet light.

At the lower levels of the ionosphere, where the intensity of the ultra-violet light has been reduced, most of the ionisation is caused by X-rays and cosmic rays which are able to penetrate further into the atmosphere. In this way an area of maximum radiation exists with the level of ionisation falling below and above it (Fig 3.3).

Often the ionosphere is thought of as a number of distinct layers. Whilst it is very convenient to think of the layers as separate, in reality this is not quite true. Each layer overlaps the others

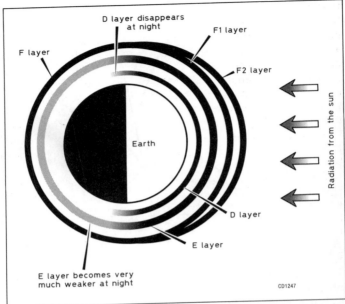

Fig 3.4. Simplified view of the layers of the ionosphere over the period of a day

with the whole of the ionosphere having some level of ionisation. The layers are best thought of as peaks in the level of ionisation. These layers are given the designations 'D', 'E', 'F_1' and 'F_2' as shown in Fig 3.4.

D layer

The *D layer* is the lowest of the layers of the ionosphere. It exists at altitudes around 60 to 90km, and is present during the day when radiation is received from the Sun. However, the density of the air at this altitude means that ions and electrons recombine relatively quickly. This means that electron levels fall after sunset and the layer effectively disappears.

This layer is typically produced as the result of X-ray and cosmic ray ionisation. It has the effect of attenuating many of the signals that pass through it.

E layer

Beyond the D layer, the *E layer* is found. This exists at an altitude of between 100 and 125km. Instead of acting chiefly as an attenuator, this layer reflects radio signals although they still undergo some attenuation.

In view of its altitude and the density of the air, electrons and positive ions recombine relatively quickly. This occurs at a rate of about four times that of the F layers which are higher up where the air is less dense. This means that after nightfall the layer virtually disappears although there is still some residual ionisation, especially in the years around the sunspot maximum which will be discussed later.

There are a number of methods by which the ionisation in this layer is generated. It depends on factors including the altitude within the layer, the state of the Sun and the latitude. However, a large amount of the ionisation is produced by X-rays and ultraviolet light, especially that with very short wavelengths.

F layer

The F layer is the most important region for long-distance HF communications. During the day it splits into two separate layers as can be seen from Fig 3.4. These are called the F_1 *layer* and the F_2 *layer*, the F_1 layer being the lower of the two. At night these two merge to give one layer called the *F layer*. The altitudes of the layers vary considerably with the time of day, season and the state of the Sun. Typically in summer the F_1 layer may be around 300km with the F_2 layer at about 400km or even higher. In winter these figures may be reduced to about 300km and 200km. Then at night the F layer is generally around 250 to 300km. Like the D and E layers, the level of ionisation falls at night but, in view of the much lower air density, the ions and electrons combine much more slowly and the F layer decays much less. Accordingly it is able to support communications, although changes are experienced because of the lessening of the ionisation levels. The figures for the altitude of the F layers are far more variable than those for the lower layers. They change greatly with the time of day, the season and the state of the Sun. As a result the figures which are given must only be taken as an approximate guide.

Most of the ionisation in this region of the ionosphere is caused by ultra-violet light, both in the middle of the UV spectrum and those portions with very short wavelengths.

Beyond the ionosphere

Above the F layer the levels of ionisation start to fall. This is mainly due to the fact that the air density is very much lower at these altitudes. As a result the ionisation is at such a low level that it has little effect on radio signals. Above altitudes of 600km the the level of ionisation is so low that it has no noticeable effects on radio communication.

Further out the Earth's magnetic field is of great importance. It protects the Earth from the *solar wind*, a steady flow of plasma or ionised particles that is emitted from the Sun. The effect of this field is to protect the Earth from this flow but its shape is modified by the plasma flow and a bow wave is created as the plasma is deflected. In times of intense solar activity these areas become overloaded and ionised particles enter the atmosphere to create an aurora as described in Chapter 5.

VARIATION IN LEVEL OF IONISATION

The level of ionisation is constantly changing. The radiation received causes ions and electrons to be released. The reverse effect also takes place. The free electrons and ions often meet, and as dissimilar charges attract one another they can combine. This means that there is a two-way process where radiation causes the electrons and ions to be generated, and the natural attraction of the two constituents means they recombine to form stable molecules again.

This all means that the level of ionisation in the ionosphere will vary. It is dependent upon a number of factors including the time of day, the season and the state of the Sun. It is not always easy to predict the exact ionisation levels, although estimates can be made giving an idea of what propagation may be possible. An understanding of what causes the variations enables the enthusiast to have an idea of what may be expected.

Variations during the day

The level of free electrons is very dependent upon the radiation being received. It is for this reason that levels of ionisation are highest in the day, falling significantly at night. Some layers even disappear at night so that they have little or no effect on radio signals. Normally the rate at which recombination takes place is slower than the ionisation process. Levels of ionisation rise swiftly in the morning as the radiation starts to be received, and then fall slowly at dusk and after dark.

Seasonal changes

The amount of heat received by areas of the Earth varies according to the season. The amount of radiation received by the ionosphere also varies although there are a few differences.

The ionisation of the E layer follows a cyclical pattern and is dependent upon the elevation of the Sun in the sky. In summer the levels of ionisation are higher, and they fall in the winter.

The F_1 layer follows a similar pattern. For most of the year it follows a similar pattern to the E layer, having a maximum level of ionisation in the summer, and being lower in the spring and autumn. In all but equatorial regions it generally merges with the F_2 layer in winter.

The behaviour of the F_2 layer is more complicated but is shows the same general trend of higher levels of ionisation in summer than the winter.

Geographical considerations

The amount of radiation that strikes the ionosphere is dependent upon the latitude in the same way that the amount of heat the Sun gives is dependent upon latitude. This means that levels of ionisation are greater in the equatorial regions and lower at the poles.

This is broadly true for the D, E and F_1 layers. However, the F_2 layer is the furthest away from the Earth's surface and is influenced by other factors. As a result other factors, including the Earth's magnetic field, have an effect and it also receives ionisation from other sources. It is found that the levels of ionisation are higher around Asia and Australia than they are over the western hemisphere including Europe, North America and Africa.

Changes on the Sun

The Sun is the source of the radiation which creates the ionosphere. As a result the state of the Sun has a great bearing upon radio conditions at any one time. Its state superimposes large changes onto the regular cycles caused by the time of day and the season.

The changes which affect the ionosphere are mainly related to the sunspots which appear on the Sun. At times of high sunspot number it is found that ionisation levels are higher and consequently radio conditions are better. Solar flares also affect the ionosphere. However, unlike sunspots which improve conditions, solar flares have the opposite effect, often causing blackouts on the HF bands. Solar flares and ionospheric storms are outlined in Chapter 5.

SUNSPOTS

One of the most notable phenomena on the Sun's surface is *sunspots*. They play a major role in determining the state of the ionosphere.

The spots themselves appear on the surface of the Sun as dark (relatively speaking) spots which can be seen if an image of the Sun is projected onto a screen or paper. The spots may last for only a few days or they may last for several months. They appear dark as the temperature on the surface at these points is only about 3000K against 6000K for the rest of the surface. The internal layers of the Sun are very much hotter than this, reaching temperatures of up to 1,500,000K.

Under no circumstances should the surface of the Sun be viewed directly, even through dark glasses. People have had their sight damaged or have even been blinded by doing this.

The spots are thought to arise from very intense magnetic fields which exist below the surface of the Sun. These fields change during the course of the cycle. At the solar spot minimum the magnetic fields are longitudinal, running from the Sun's north to its south. As the Sun rotates it is found that the fields rotate at a different speed, the equatorial regions spinning more slowly than the poles. This causes distortion of the fields which slowly align in an east/west direction. It is found that as a result of these changes the activity on the Sun changes in a cyclical form, having a distinct trough and peak.

At times the changes in the magnetic fields cause eruptions to take place through the Sun's surface. Around the eruptions the surface temperature falls dramatically, giving rise to what are seen as dark spots.

The area around the spot has a large temperature differential, and there is also a large disturbance in the magnetic field in this area. The magnetic fields are unusually strong and this results in enormous forces being produced. It is also found that large levels of electromagnetic energy and cosmic rays emanate from around these spots and this affects the levels of ionisation in the ionosphere.

Sunspots often appear in clusters. Each spot can be anywhere between a few hundred kilometres in diameter to large ones which may be almost 150,000km wide. The groups themselves can contain several large spots and may be over 400,000km in diameter.

It is of interest to note that apart from the dark sunspots which appear, there are also bright areas called *plages*. These are associated with the sunspots, but comparatively long lived. The plages appear first, and within them the sunspots appear. It is also found that the plages remain after the sunspots have disappeared.

Sunspot numbers

It can be imagined that different astronomers with different equipment will count different numbers of sunspots. Shortly after the discovery of sunspots a scientist named Rudolf Wolf, who was the director of the Zurich observatory, devised a system giving the relative sunspot activity. This could then be used as a common standard for all observatories. The activity is defined in terms of a *sunspot number*, which is not the actual number of sunspots which are seen.

$$R = k(10g + s)$$

where R is the Wolf number for sunspot activity, k is a correction factor to take account of the equipment and observer characteristics, g is the number of sunspot groups and s is the number of observable spots (whether individually or in groups).

This formula weights the sunspot activity number heavily towards clusters. This may appear to be incorrect, but Wolf chose to adopt this system because he deduced that the large clusters were a better sign of sunspot activity than short-lived spots on their own. Despite its shortcomings it is still a very useful indicator because sunspot numbers have been measured in this way for over 200 years and it gives very useful comparisons and visibility of trends over the whole period.

The day-to-day numbers of sunspot activity vary widely. In order to view the trends the data is averaged or smoothed over a wider period. Two stages are used for this. First the daily numbers are averaged over the period of a month and then the monthly figures are smoothed over a 12-month period. In order to ensure that the mean falls right in the middle of the month in question rather than between months, the period of smoothing is run over 13 months, but taking half the value for the months at either end:

$$R_s = \frac{\frac{1}{2}R_{m1} + R_{m2} + R_{m3} + \ldots + R_{m11} + R_{m12} + \frac{1}{2}R_{m13}}{12}$$

where R_s is the smoothed sunspot number and R_{m1} to R_{m13} are the monthly averaged numbers for months 1 to 13.

Both the monthly average sunspot numbers and the smoothed values are available for use in propagation predictions, although the smoothed figures are much in arrears. The numbers are now prepared by the Sunspot Index Data Centre in Brussels from information supplied by a number of observatories. They appear in DX propagation information available from a wide variety of sources including the RSGB. The 12-month smoothed sunspot number correlates quite closely with the prevailing HF radio propagation conditions.

Sunspot cycle

Although the numbers of sunspots had been observed and logged since 1749, it was not until about a hundred years later that it was discovered that there was a regular variation in the sunspot number. The reason it had laid undetected for so long was due to the fact that the daily numbers varied so widely, and no trends could be seen amidst the wild variations.

Once the numbers were smoothed a trend was detected and it could be seen that there was cyclical variation with a period of about 11 years. These cycles are now given numbers starting with Cycle 1 which began in 1755. Since then there have been over 20 more with Cycle 22 ending in the latter part of summer 1996.

From analysis of the figures since the first records were available a number of factors have become obvious. The first is that

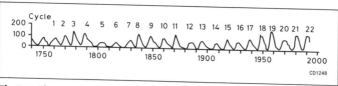

Fig 3.5. The sunspot cycles since records began

the cycles are by no means very regular. Although the average cycle length when measured as the time between two peaks is about 10.9 years, it is found that it varies anywhere from just over seven years to 17 years. The smoothed sunspot numbers also vary widely. The maximum sunspot activity numbers vary from 49 to 200 with an average of just over 100. The minimum number can be anywhere between none and 12.

Usually it is found that sunspot activity rises sharply after the sunspot minimum, reaching the peak in around four years, and after this it falls away more slowly, taking around seven years to decay. Naturally this figure too varies very widely and it can only be taken as a very rough guide.

The sunspot cycle is of great interest to anyone using the HF portion of the radio spectrum. Propagation conditions are greatly influenced by sunspot activity, and accordingly they vary in line with the sunspot cycle. At the low point of the cycle, the high frequency bands above 20MHz or so may not support ionospheric reflections, whereas at the peak of the cycle frequencies at 50MHz and higher may be reflected.

Ionospheric indicators

The number of sunspots which are seen at any time gives an indication of the amount of radiation being received by the ionosphere, and hence the level of ionisation. Although the sunspot number can be calculated, a more objective method of achieving this is to look at the amount of radiation coming from the Sun. It is found that not only does the Sun emit vast quantities of heat and light, it also emits energy at radio frequencies. An indication of solar activity can be made by measuring the level of solar noise. This is normally taken at a frequency of 2800MHz (a wavelength of 10.7cm) and is called the *solar flux*. It has been shown that this

Table 3.1. Conversion between A and K indices

K index	0	1	2	3	4	5	6	7	8	9
A index	0	3	7	15	27	48	80	140	240	400

follows the daily sunspot number and a number of relationships have been devised. One of the most straightforward is:

$$\text{Solar flux} = 73.4 + 0.62R$$

where R is the daily sunspot count.

At times the Earth's magnetic field is affected by solar activity. These geomagnetic disturbances often result from bombardment of low-energy particles from the Sun. In turn this can result in an ionospheric storm which can disrupt communications on the HF bands. It is given in the form of two indices. The first is called the *K index*. This is measured on a scale between 0 and 9 where 0 is the quietest and 9 indicates the most disturbed. It is based on values measured at several points around the Earth and is updated every three hours.

Another index called the *A index* is also available. This can vary between 0 and 400. However, the A index is in a 24-hour form and is therefore less immediate than the K index. A conversion between the two indices is given in Table 3.1.

Chapter 4
IONOSPHERIC PROPAGATION

Propagation using the ionosphere is one of the most important means of long-distance communication around the globe. Whilst satellites are being used increasingly for many applications, ionospheric propagation is still very important and will always remain so. It has many advantages, being cheap to operate, unlike satellites which are very expensive, and using the ionosphere it is easy to set up a link between two points. Broadcasters widely use this means of propagation for transmissions to other countries. Radio amateurs have long been keen to utilise the opportunities the short-wave bands offer in terms of long-distance contacts, and many commercial users still find it the only viable way of establishing long-distance communications.

To make the best of ionospheric propagation it is necessary to know the ways in which signals propagate when the ionosphere is used.

GROUND WAVE

On low and medium frequencies a form of propagation known as *ground wave* is used. The signal spreads out from the transmitter along the surface of the Earth. Instead of just travelling in a straight line the signals tend to follow the curvature of the Earth. This is because currents are induced in the Earth's surface and this action slows down the wave-front in this region, causing the wave-front to tilt downwards towards the Earth as shown in Fig 4.1. With the wave-front tilted in this direction it is able to curve around the Earth and be received well beyond the horizon.

The ground wave is generally used for signals below about 2 or 3MHz. The reason for this is that the level of attenuation rises with increasing frequency. A short-wave broadcast station will only be audible for short distances via the ground wave. In comparison, medium-wave stations are audible over much greater distances – typically a high-power station may have a coverage area of a hundred kilometres or more whilst low-power stations may be heard for distances of a few tens of kilometres.

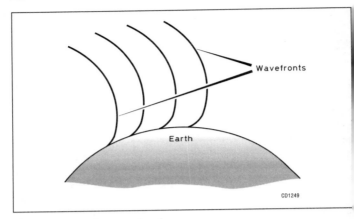

Fig 4.1. A ground-wave signal

There are many factors which affect the distances over which ground-wave signals can be heard. Apart from the transmitter power, the terrain over which the signal travels has a major effect. Sea water gives the best performance. It is better than moist soil and that in turn is better than dry soil. Desert and rocky terrain give very poor performance and this means that high powers and good antennas are needed for these areas.

The type of antenna has a major effect. Vertical antennas are used because it is found that the attenuation of a vertically polarised signal is very much less than a horizontally polarised one. In some cases the difference can amount to several tens of decibels. It is for this reason that medium-wave broadcast stations use vertical antennas, even if they have to be made physically short by adding inductive loading.

D LAYER ATTENUATION

As far as radio signals are concerned this layer acts as an attenuator. The attenuation varies as the inverse square of the frequency, so doubling the frequency reduces the level of attenuation by a factor of four. This means that low-frequency signals are prevented from reaching the higher layers, except at night when the layer disappears.

The signals are attenuated because the radio signals excite the free electrons into vibration. When they vibrate they collide with other molecules, and at each collision a small amount of energy is lost. This means that the level of attenuation is dependent upon the number of collisions which occur. In turn this is dependent upon a number of other factors.

The first is the number of gas molecules present. In the region of the D layer this is relatively high, and this is why this layer rather than those at a greater altitude where the air density is less causes signals to be attenuated. Another factor is the level of ionisation. Only when there is a sufficient level of ionisation does this layer give rise to attenuation. At night when the level of ionisation falls, so does the degree of attenuation. Thirdly it depends upon the frequency in use. As the frequency increases, the wavelength shortens, and the number of collisions between the free electrons and gas molecules decreases. It is for this reason that signals lower in the frequency spectrum are attenuated far more than those which are higher in frequency. Even so high-frequency signals still suffer some reduction in signal strength.

The level of absorption reduces the signal level sufficiently to prevent signals on the lower frequencies from passing through. Typically signals on the medium waveband are totally attenuated by the D layer during the day. This means that signals are only audible over the distances reached by the ground wave. At night when the D layer disappears signals reach the higher layers and can be heard over much greater distances. At higher frequencies the D layer still causes signals to be attenuated.

E AND F LAYER REFLECTION

When a signal enters the E or F layers it excites the electrons into oscillation as in the D layer. However, the air density is much less and the number of collisions is much reduced. As a result the predominant effect is that the electrons re-radiate the signal. As the signal is travelling in a medium where the density of electrons is increasing the signal is refracted away from the area of higher electron density.

This refraction is often sufficient in the case of HF signals to send the signals back to the Earth. In effect it appears that the

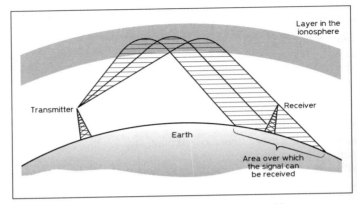

Fig 4.2. Refraction of a signal as it enters an ionised layer

signal has been reflected by the layer. The tendency for this reflection is dependent upon the frequency and the angle of incidence. As the frequency increases, so the amount of refraction decreases until a frequency is reached where the signals pass through the layer and on to the next. Eventually a point is reached where the signal passes through all the layers and on into outer space.

During the day it is found that signals in the medium-wave band only propagate using the ground wave. If the frequency of the signal is increased a point is reached where the attenuation caused by the D layer become sufficiently small to allow a signal through to the E layer. Here it is reflected and will again pass through the D layer to return to the ground where it may be heard at a considerable distance from the transmitter.

If the frequency is increased further the effect of the refraction at the E layer reduces to a point where it will pass through and reach the F_1 layer. Here it may be reflected and pass through the D and E layers before reaching the ground again. As the F_1 layer is higher than the E layer the distance reached will be greater than that for an E layer reflection.

Again, as the frequency rises the signal will eventually pass through the F_1 layer and onto the F_2 layer. As this is the highest of the layers the distance reached using this is the greatest. Typically

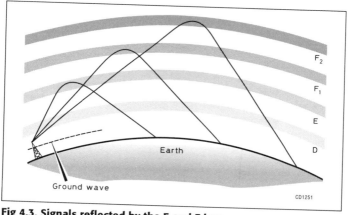

Fig 4.3. Signals reflected by the E and F layers

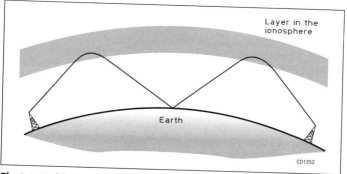

Fig 4.4. Multiple reflections

the maximum skip distance for the E layer is around 2500km and 5000km for the F_2 layer.

Whilst the distances which can be achieved by reflections from the F_2 layer are very considerable, they do not explain why signals can travel to the other side of the globe. This occurs because signals undergo more than one reflection. Having returned to the ground after the first reflection, the signals are reflected by it back up to the ionosphere again, and then back to ground as shown in Fig 4.4. Naturally each reflection introduces more losses and

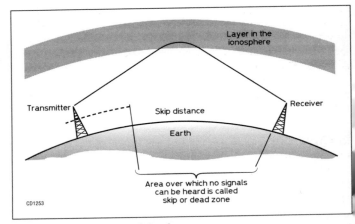

Fig 4.5. Skip distance and dead zone

therefore the signals are attenuated each time. The surface of the Earth at the point of reflection also has an effect. Sea water is a very good reflector as might be expected, but dry desert is very poor. This means that signals which are reflected in the Atlantic are likely to be stronger than those reflected by a desert region.

In addition to the losses at each reflection signals are attenuated each time they pass through the D layer. These losses normally constitute a major loss in the signal path, particularly as the signal has to pass through the D layer twice for each reflection by the ionosphere. As already mentioned this loss reduces with frequency. Apart from the fact that high-frequency paths are more likely to use the F_2 layer and have less reflections, the high-frequency path will also suffer less loss from the D layer and it will mean that a signal on 28MHz for example will be stronger than one on 14MHz if propagation can be supported at both frequencies.

SKIP DISTANCE AND SKIP ZONE

When a signal travels to the ionosphere and is reflected back to ground, the distance which it travels is called the *skip distance* as shown in Fig 4.5.

The ground-wave signal will be only heard for a certain

distance from the transmitter, due to the attenuation of the signal. The signals travelling to the ionosphere may not be reflected until they reach well beyond the place where the ground-wave signals fade out. As a result there will be an area or zone where no signal is heard. This is known as the *skip zone* or *dead zone*. This is particularly pronounced for high-frequency signals where the ground wave fades away quickly and the skip distance may be 1000km or more.

CRITICAL FREQUENCY

One of the parameters of the ionosphere which can be measured is called the *critical frequency*. This is measured by sending a pulse directly upwards. Initially it will be reflected and can be received at the point of the transmitter. The time taken for the pulse to be returned can be measured and the effective height of the layer can be measured. However, as the frequency is increased a point is reached where the pulse passes straight through the layer and on to the next, or into outer space depending upon the layer in question. The frequency where this occurs is called the *critical frequency* for that layer. Critical frequencies can be measured for each layer and these can be used to help determine the state of the ionosphere.

LUF

As the frequency of a transmission is reduced losses increase. This arises from increased attenuation from the D layer and the fact that further reflections may be needed. This means that for any given path there is a lowest frequency which can be used. The *LUF* (lowest usable frequency) is defined as the frequency at which the signal equals the minimum signal strength required for satisfactory reception.

From this it can be seen that the LUF is dependent upon the stations at either end of the path, their powers, the antennas and the levels of noise in the area of the receiver. Other stations nearby may have a different LUF because their characteristics may be different. The LUF is also dependent upon the type of modulation being used. Morse code, which only requires a narrow bandwidth and can be copied at a lower signal strength, will have a lower LUF than a station using SSB for which a wider bandwidth and higher signal strength is required.

If it is necessary to use a frequency below the LUF then, as a rough guide, it is found that by increasing the powers or antenna gains by a factor of 10dB the LUF can be lowered by about 2MHz.

It is also necessary to note that the LUF *increases* in periods of high solar activity. This is because the increased radiation from the Sun increases the D layer ionisation and hence its absorption. This means that at the peak of the sunspot cycle there is a degradation in the low-frequency bands.

MUF

When a signal is transmitted over a certain path there is a maximum frequency which can be used. This results from the fact that as the transmission frequency is increased the signal penetrates further into the ionospheric layers until it passes straight through. When the point is reached where communications can just start to fail this is known as the *MUF* (maximum usable frequency). This is generally three to five times the critical frequency, dependent upon the layer being used and the angle of incidence of the transmission.

ANGLE OF RADIATION

The angle at which the signal leaves the antenna and travels towards the ionosphere is important. For this the angle of radiation is defined as the angle between the main lobe of the signal and the ground. Geometry shows that if the angle of radiation is low then the signal will travel further before it is reflected, and hence the total distance travelled will be further (Fig 4.6).

Even a relatively small increase in the angle of radiation can considerably reduce the distances covered. The maximum distance which can generally be achieved using a reflection from the E layer is considered to be 2500km. This is reduced to just 400km if the angle of radiation is 20°. Similarly the distances using the F_2 layer are reduced from 5000km to 1000km. However, it should be noted that signals with a low angle of radiation travel further through the D layer than those with a high angle. This means that they will suffer greater amounts of attenuation. For this reason low-angle signals are generally best received on

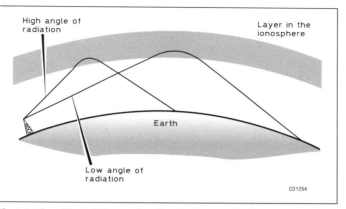

Fig 4.6. The effect of the angle of radiation on the distance travelled

high-frequency bands where attenuation by the D layer is considerably less. Night time also brings reductions in D layer attenuation, allowing lower-angle signals to be received.

In targeting their transmissions, broadcasters make use of both the direction or azimuth of the signal and the angle of radiation or elevation. In this way they are able to ensure the maximum signal reaches the required target area. For radio amateurs who are mainly interested in achieving the maximum distance it is usual to ensure that the antenna has a low angle of radiation.

OPTIMUM FREQUENCIES

In order to receive or send signals to a given location, there are likely to be a number of different paths which can be used. Sometimes it may be possible for the signals to travel via both the E and F layers, or one path may use two hops instead of one. In fact the picture is rarely as well defined as it may appear to be from the textbooks. However, it is still possible to choose a frequency from a variety of options to help improve the possibility of making contact with a certain area.

In general higher frequencies give the best results. This is because the effect of attenuation by the D layer is reduced. Although signals may pass through the D layer they still suffer some attenuation as mentioned already. Increasing the frequency also means that the higher layers are used. As losses are incurred at each reflection, either at the ionosphere or at the Earth, the minimum number of reflections is desirable. In some cases it may be possible to use a lower frequency and two hops via the E layer, or a higher frequency and one hop using the F layer. In this case a single hop would be preferable.

When using the higher frequencies it is necessary to ensure that communications are still reliable. In view of the ever-changing nature of the ionosphere a general rule of thumb is to use a frequency which is about 20% below the MUF. In most cases this will ensure that most expected changes can be absorbed. Obviously the path will open and close at different times of the day, and therefore it is necessary to take account of this as well.

FADING

One of the major features about signals which are received via ionospheric propagation is that the signals fade in and out. Often this can result in the signal changing in strength by many tens of decibels. There are a number of reasons for this, but they all result from the fact that the state of the ionosphere is constantly changing.

The most common form of fading results from *multipath interference*. This results from the fact that a signal reaches the ionosphere over a wide area because the beamwidth of the antenna is finite. As the ionosphere is very irregular the signal will reach the receiver via a variety of paths (Fig 4.7). This means that the received signal is the summation of a number of signals which have travelled over slightly different paths. As the ionosphere changes, the signals will fall in and out of phase with one another, resulting in the strength varying by a considerable degree.

Another time when fading occurs is at night on low-frequency signals using the ground wave. Under daytime conditions only the ground-wave signal is audible. At night when the D layer disappears signals are also heard via the sky wave. The changes in the ionosphere will cause the sky-wave and ground-wave signals to fall in and out of phase with one another, causing fading and distortion. This can occur even at distances a few kilometres from the transmitter.

In some instances the receiver may be on the edge of the skip

IONOSPHERIC PROPAGATION

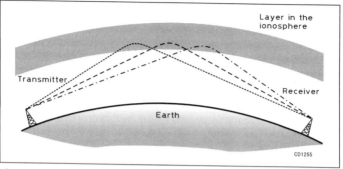

Fig 4.7. A signal is received via several paths

zone. In this case the changes in the ionosphere may cause it to pass in and out of the skip zone, resulting in large changes in signal strength.

Another form of fading results from changes in polarisation. When the ionosphere reflects signals back to ground they may be in any polarisation. For the best reception the polarisation of the antenna and the incoming signals should be the same. However, as the polarisation of the signals changes, matching and then not matching that of the antenna, this will cause variations in signal strength.

GREY LINE PROPAGATION

It is found that propagation where signals which travel around the globe along the *grey line* or *twilight zone* where dusk is falling or dawn is breaking is extremely efficient. Signals which travel in this zone can be heard at remarkable strengths because path losses are much lower. In the UK it is not unusual to hear signals on 3.5MHz from New Zealand, which is at the other side of the globe, at the same strengths as many local stations. Similarly on the higher-frequency bands signals from low-powered stations or those with poor antenna systems can be heard.

The reason for this enhanced propagation is that at this time of day the ionisation in the D layer is very much reduced or virtually non-existent. Either the level of ionisation is falling at dusk, or it is still to build up at dawn. The effect of the D layer in attenuating signals is correspondingly reduced. In addition to this the ionisation in the F_2 layer decays more slowly at night and being higher in altitude the level of radiation does not decay as quickly. Similarly, at dawn the F_2 layer ionisation is still at a relatively high level and its level of ionisation starts to rise more quickly than that of the D layer. These effects mean that at dawn and dusk the absorption by the D layer is minimal, whilst the ionisation in the F layer has not fallen by a large degree. This gives the opportunity for signals to travel over great distances with relatively low levels of attenuation.

Chapter 5
IONOSPHERIC STORMS AND AURORAS

The conditions on the short-wave bands are always changing. Sometimes a complete or virtually complete fade-out occurs when no HF communications are possible via the ionosphere. It is possible for VHF operators living far enough north or south to communicate over long distances at these times, and there may even be a visible aurora. All these events are fascinating, and the visible aurora is a magnificent sight to see. The origin of all these phenomena can be traced back to the Sun and the events which occur there.

The way in which ionospheric disturbances and auroras occur is quite complicated. A number of effects are linked together, and many of them occur in the outer reaches of the atmosphere and beyond.

It was only after the first rockets were launched that real measurements could be made, and even now there is a considerable amount of work to be undertaken before a full understanding is achieved. As a result sometimes there are different interpretations of how exactly how the auroras occur. Nevertheless there is a large amount which is known and understood about these phenomena.

SOLAR WIND

Apart from providing heat and light the Sun also emits vast amounts of particles. This stream of material is called the *plasma* and it consists mainly of hydrogen and helium molecules. However, due to the extreme temperatures on the surface of the Sun these molecules have become ionised to give ions, protons, neutrons and electrons. As a result this plasma is conductive and is affected by any magnetic fields which are encountered although its overall charge is zero.

This stream of plasma (the *solar wind*) is always present and travels exceedingly quickly, usually between 300 and 400km per second. However, the density is relatively low with only around five ions per cubic centimetre. Both these figures can rise

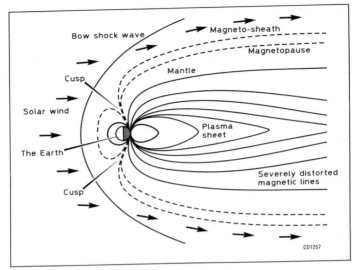

Fig 5.1. The Earth's magnetic field

dramatically during disturbances on the Sun, when speeds of up to 700km/s and densities of 80 ions/cm^3 can be detected.

Outer reaches of the atmosphere

Under quiet conditions the outer reaches of the Earth's atmosphere are able to deflect the solar wind. The ionosphere which is used for short-wave communications extends to an altitude of around 400km.

Above this is an area called the *magnetosphere* where the Earth's magnetic field extends. This field extends many thousands of kilometres out from the Earth, although it is greatly influenced by a number of factors in outer space. One of the main influences is the solar wind. This compresses the field on the sunward side of the Earth, whilst it becomes distended on the other side as shown in Fig 5.1.

In general the magnetic field prevents particles from the solar wind from entering the Earth's atmosphere, although some can enter through cusps in the polar regions.

IONOSPHERIC STORMS AND AURORAS

SOLAR FLARES

On occasions normal propagation conditions on the HF bands are disrupted. Sometimes it may be that communications are disrupted for a few hours whilst other times conditions may be affected for several days. There are a number of reasons for these disruptions but one of the main causes are *flares* which occur on the Sun.

These solar flares are often associated with sunspots and as a result they occur more often around the periods of high sunspot activity. The flare results when very high magnetic fields cause an eruption to take place though the solar surface and material is ejected. As the eruption takes place it causes the area around the sunspot to become intensely heated as material from under the surface of the Sun rises.

There are a number of stages in a flare. First a large amount of very-high-energy particles which are mainly protons are ejected. These travel at about a third the speed of light. Simultaneously there is a rapid rise in the level of radiation at all frequencies including ultra-violet, X-rays and cosmic rays as well as radio frequency emissions. The next stage of the flare occurs after about five minutes and constitutes the emission of lower-energy particles. These only travel at a speed of around 1000km per second. The particles which are ejected may or may not be on a trajectory which intersects with the Earth. This means that it is quite possible that the only effects which will be noticed are those of the increase in level of radiation.

The whole flare only lasts about an hour, after which the Sun's magnetic fields start to subside. However, the effects are soon felt by the Earth (Fig 5.2).

SIDs

The first effect to reach the Earth is the electromagnetic radiation which arrives just over eight minutes after the flare starts. As the radiation received includes ultra-violet as well as X-rays there is a rapid increase in the level of ionisation, particularly in the D layer.

This causes the level of absorption to increase, affecting the lower bands first and progressively the higher ones as the level increases. This can cause a complete blackout of the short-wave

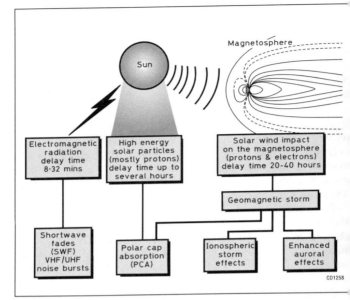

Fig 5.2. Effects on the ionosphere that are caused by solar flares

bands in some instances, whereas others may only affect the lower bands.

These conditions may last up to a couple of hours with the high-frequency bands recovering first. Also it is only the sunlight side of the Earth which is affected, so any areas which are in darkness when the flare occurred will escape the effects, assuming the increased level of radiation has subsided.

Very small flares will not increase the level of D layer ionisation to such an extent that all the HF bands close. However, there is an overall increase in the level of ionisation at all layers and this may improve the conditions in the higher-frequency bands for a while. It is also found that several small flares may occur improving HF conditions before a major flare occurs bringing significant reduction in conditions.

In view of the speed with which these changes happen they are called *sudden ionospheric disturbances* (SIDs), *short-wave fades*

SWFs) or the *Dellinger effect*. At the same time as the X-ray radiation reaches the Earth, causing a SID, noise bursts from the increased levels of radio frequency radiation may be monitored VHF.

Polar cap absorption

It takes a few hours for the high-energy particles to arrive. When they do the Earth's magnetic field prevents them from entering all around the atmosphere. Instead they move to the poles and enter at these points. Here they cause a very large increase in the level of D layer ionisation, typically increasing absorption by between 40 and 80dB. This is known as *polar cap absorption* (PCA) and these conditions manifest themselves suddenly and may last for up to three or four days, or in the cases of a large event for a little more. They have the effect of totally blacking out radio communications in the polar regions including signals which pass across them. However, users of the radio spectrum in other parts of the globe may not be aware of the blackout except that no signals are heard via polar paths.

Geomagnetic and ionospheric storms

Between 20 and 40 hours after the flare the lower-energy particles arrive. These considerably increase the general level and velocity of the solar wind. When this arrives it meets the Earth's magnetic field, causing changes in magnetic activity. The level of magnetic activity can be recorded by the A or K indices which also act as an indicator of the stability of short-wave conditions.

The occurrence of a *geomagnetic storm* may in turn give rise to an ionospheric storm, although this does not happen every time. It is also worth noting that an ionospheric storm cannot be present without the presence of a geomagnetic storm. The *ionospheric storm* affects the chemistry of the ionosphere, depressing the levels of ionisation by a significant degree. This results in the frequencies which can be reflected by the F layer being reduced by half or more – the E layer is also affected in a similar way. There is also an increased level of absorption. This means that the maximum usable frequency falls as a result of the depleted F layer, and the lowest usable frequency rises as a result of the increased absorption. In turn, this results in the band of frequencies which can be used over a given path being reduced. If the two frequencies (LUF and MUF) meet then there is a complete radio blackout and no long-distance communications are possible over that path. Indeed, during a bad storm it is possible that no stations may be heard over the whole of the HF spectrum except for local stations heard via ground wave.

The beginning of an ionospheric storm usually begins around the polar regions and then spreads out to affect the middle and low latitudes towards the Equator. It also occurs on both the light and dark areas of the world at the same time, unlike SIDs which only affect the daylight areas. Unlike the SID it develops gradually, it lasts longer (periods of a week are not unknown, although a few days is more normal) and then conditions slowly return to normal.

Even after the bands have recovered the effects of the flare may be noticed some time later. The Sun has a period of rotation of about 27 days. This means that when it sweeps columns of particles out into space there is a tendency for them to hit the Earth again about a month after the first occurrence. As a result these ionospheric storms often recur at this interval, but usually with a diminished intensity.

VISIBLE AURORAS

Other effects can be noticed at these times. One is that visual auroras can be seen around the poles. These are exceedingly beautiful, and consist of coloured glows changing in the sky. The main colours are whites, greens and reds resulting from the spectral lines of atomic oxygen and nitrogen. On some occasions bluish tints can also be seen. Sometimes the display can consist of streamers of colour which change their shape and colour over a period of a few minutes. These displays are caused by free electrons from the increased solar wind entering the atmosphere around the magnetic poles. As the electrons descend they collide with other molecules in the upper atmosphere and release energy, some of which is light.

Auroral propagation

When the density of electrons entering the atmosphere at the poles is sufficiently high this can have a significant effect on the fre

IONOSPHERIC STORMS AND AURORAS

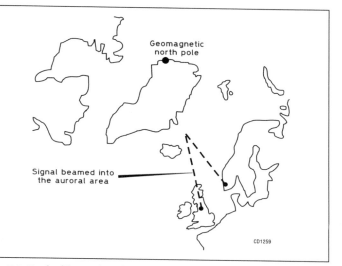

Fig 5.3. Typical beam headings when using auroral propagation

electron density in these areas. When this happens it can bring the levels of ionisation in the E layer to sufficient levels to reflect signals at VHF despite the fact that ionisation levels elsewhere are depressed. Frequencies affected vary from one aurora to the next. It has been known for reflections to be detected at frequencies over 1000MHz although the more usual maximum is around 450MHz. Propagation via this mode can only occur relatively close to the poles. Generally the latitudes of the stations must be higher than about 55°, although when there are large auroras it can extend further away from the poles. Stations in Italy have been reported on rare occasions.

When making contact via this mode of propagation antennas must be beamed towards the auroral zone, and signals are normally heard using back scatter. This means that the optimum beam heading will be in a different direction to that of the station being heard or contacted. In fact the direct path is far more likely to be off the back or side of the beam as shown in Fig 5.3.

It is found that a signal reflected by auroral ionisation is distorted badly. Signals are subject to rapid flutter because of the changing state of the ionosphere. The flutter can be very rapid, appearing as a rough low-frequency buzz with a frequency of up to 50 or 60Hz. The buzz varies considerably and changes in nature from one auroral event to another and even during the course of an aurora. It makes speech transmissions very difficult to copy, although SSB can be used on occasions. CW is much better and can be copied relatively easily despite the distortion.

In addition to the flutter a Doppler shift is added to signals which are reflected. This is caused by the electrons which are entering the atmosphere and moving towards the Earth. As they act as minute points of reflection and they are moving, this adds a shift to any signals. The electrons are not moving uniformly towards the Earth and this means that there is a spread in the Doppler shift which is placed onto the signal, manifesting as a distinctive hissing sound. The centre of the Doppler shift is dependent upon the frequency in use. For 144MHz, which is the most popular band for this mode of communication, it gives a shift of up to 500Hz.

Chapter 6
PREDICTING, ASSESSING AND USING HF PROPAGATION

It is very important for a whole variety of users of the HF portion of the radio spectrum to be able to assess and predict the propagation conditions. For example, broadcasters need to know which frequencies will enable them to reach their target areas best. This has to be planned months in advance so that frequencies can be arranged and new service schedules can be determined. Other users as well need to know what the state of the ionosphere is likely to be or what it is at that time. For radio amateurs and short-wave listeners the conditions are of particular importance because they enable estimates to be made of when to listen for particular areas. If a beam is used, it enables it to be directed to areas where propagation conditions may be best supported. These and many other reasons make estimates of propagation of great interest.

To cater for all these needs a variety of techniques are used, and a science akin to weather forecasting has been built up. Using a variety of measurements it is possible to detect the state of the ionosphere at any given time, and using various other methods it is possible to predict propagation conditions in the future.

KNOWLEDGE OF THE BANDS
One of the best ways of knowing what to expect on the different bands, or knowing where to listen to receive stations from certain areas, is from experience. After listening on the bands for a while it is possible to gain a good idea of what to expect. A broad overview of the bands up to 30MHz is given in Table 6.1.

AMATEUR BANDS
There are several amateur bands within the MF and HF portions of the radio spectrum. Although they are covered by the broad summary in Table 6.1, each one has its own character and is likely to produce signals from different areas and at different times.

Table 6.1. Overview of propagation modes below 30MHz

Frequency band	Modes of propagation
30–300kHz (LF)	Propagation is from ground wave and reflections from the lower ionosphere. High powers and large antennas are normally required. Fading often occurs where both ground-wave and ionospheric reflections are present.
300kHz–3MHz (MF)	Ground-wave propagation only during the daytime. At night there is propagation from the weakening E layer. Ranges beyond the ground wave are only possible at night.
3–30MHz (HF)	Short distances obtained via E layer, but virtually all long distances via F2 layer. Daytime attenuation is caused by D layer, particularly at low frequencies. Propagation is variable and dependent upon time of day, season, sunspot cycle and presence of ionospheric storms etc.

160m (Top Band) (1.81–2.00MHz)

The exact allocations for this band vary from one country to the next. That given above is correct for the UK. Much of this band is shared with other services. During the day signals are heard via the ground wave and, depending upon transmitter powers and antennas at the transmitter and receiver, distances reached may be 40km or more. At night when the D layer disappears distances increase and it can be possible to hear stations several hundreds of kilometres away. It is possible to make transatlantic contacts when conditions are right if sufficiently good antennas are available at both ends.

For very-long-distance contacts the whole of the path must lie in darkness. However, there can be a significant improvement at dawn and dusk for contacts with the other side of the globe. These enhancements may only last for 10 to 15 minutes at maximum and sometimes they may only last for a minute or so.

For shorter paths like those between Europe and North America signals peak when it is either sunrise or sunset at one end or the other. For this particular path sunset in the USA corresponds to about 0200 in the UK. Long-distance north/south paths often peak around midnight.

As a general rule long-distance work improves in winter because of the longer hours of darkness and lower levels of static. As this does not correspond with optimum conditions in the other hemisphere it means that these signals may be heard best around the spring and autumn equinoxes.

80m (3.5–3.8MHz [3.5–4.0MHz in North America])

Like Top Band this one is also shared with other services and can be noisy, especially at night. During the day the distances which can be reached are greater than those on Top Band. Often stations a few hundred kilometres away can be heard. At night stations from further afield can be heard – distances of over a thousand kilometres are common, and greater distances can be achieved by those with good antennas. The band comes into its own during the years of the sunspot minimum but it can perform well at any time.

Propagation along the grey line can produce exceedingly good results, with stations from the other side of the globe being audible at the same strengths as many local stations. However, this may only be short lived and it can be very selective in terms of location.

40m (7.0–7.1MHz [7.0–7.3MHz in North America])

The 40m band is a good band for long-distance work. During the day, stations up to distances of a few hundred kilometres can often be heard. However, ionospheric absorption limits greater distances. The high angle of skip means that the skip zone is small or non-existent.

At night distances increase considerably. Stations from further afield are more apparent and as the skip zone increases local stations fall in strength. It is a favourite band for many during the low part of the sunspot cycle, being capable of long-haul contacts during the hours of darkness. Again, the grey line can produce some spectacular results, as can openings which occur at dusk and dawn.

In Europe the band is only 100kHz wide, making it congested when open for long-distance traffic. In North America, where frequencies up to 7.3MHz are available, interference from European broadcast stations (to whom this portion is allocated in Europe) can be a problem.

30m (10.100–10.150MHz)

This band was released for amateur use after the World Administrative Radio Conference held in 1979 (WARC 79). It is still not widely used but is capable of giving good results. It is very similar in character to the 40m band, being only slightly higher in frequency.

For most of the time a skip zone is apparent, except at the peak of the sunspot cycle. The level of absorption is less than on 40m and during the night distances increase. Propagation is again best when the path lies in darkness, but again conditions are enhanced by grey line and dusk or dawn conditions.

It is also found that during periods of the sunspot minimum when ionisation levels are lower, absorption is sufficiently low to allow long-distance contacts throughout the day.

20m (14.0–14.35MHz)

This is the main long-haul band for radio amateurs, reliably giving the possibility of long-distance contacts during all phases of the sunspot cycle. During the day stations up to about 2500 or 5000km can be heard when conditions are good, and there are virtually always stations up to 2000km which can be heard. The band normally closes at night during the winter and during the sunspot minimum. During the summer and the sunspot maximum it will remain open most of the night. Spring and autumn normally produce good results with stations from the other hemisphere being heard with ease at various times of the day.

17m (18.068–18.168MHz)

Like the 30m band this one was also released for amateur use after WARC 79. It is very much a half-way house between the bands either side of it.

15m (21.0–21.45MHz)

This band is more variable than 20m, being affected more by the state of the sunspot cycle. During the peak it is open during the day and well into the night when it will support propagation over many thousands of kilometres. When the band is open like this strengths are usually better on this band than 20m because of the lesser effect of the D layer. Conditions are usually not quite s good in the early morning, improving as the day progresses. During the sunspot minimum few stations may be heard during the day and none at night.

12m (24.890–24.990MHz)

This is another of the WARC 79 bands. It is greatly affected by the position in the sunspot cycle, and has many similarities with 10m, although it will just support propagation when 10m cannot. However, it will follow very much the same pattern as the slightly higher frequency band.

10m (28.0–29.7MHz)

This is the highest-frequency band in the short-wave (HF) portion of the spectrum. During the sunspot minimum it may only support ionospheric propagation via sporadic E which occurs mainly in the summer months. This gives propagation over distances of about 1500km.

At the peak of the sunspot cycle it give excellent possibilities for long-distance contacts, producing very strong signals. This band is well known for enabling stations with low powers and poor antennas to make contacts over great distances as ionospheric absorption is less than on the lower frequency bands. In general propagation on these frequencies requires that the path is in daylight. Despite this at the peak of the sunspot cycle the band may remain open into the night, although it will eventually close.

This is very much a rough guide to the different bands and should be treated as such. However, it can be used as a starting point from which a feel for what each band can produce can be gained. This is best done by spending time listening on them to see what can be heard or contacted.

BEACONS

As propagation conditions can vary from one day to the next, beacon stations are used to indicate whether a path exists to a certain area. There are beacons on many of the bands, although the majority are on 20 and 10m.

There is a world beacon network. When complete it will have 18 stations around the world transmitting on 14.100, 18.110,

21.150, 24.930, and 28.200MHz. As they share the same frequency their transmissions are time coordinated so they do not overlap. They transmit for 10 seconds on each frequency in rotation, falling silent until their position in the next three-minute cycle.

The use of these beacons gives an excellent indication of conditions. They can be used for propagation research as well as enabling the listener or transmitting amateur to gain a good idea of the prevailing conditions.

GREY LINE PREDICTIONS

Grey line propagation is very important because it can produce some excellent results. Stations can be heard over long distances and at very good signal strengths. In order to make the best use of this form of propagation it is necessary to discover where the grey line will be. It does not remain the same all year round as the axis of the Earth relative to the Sun changes with the seasons. During the months of December and January the northern hemisphere is angled away from the Sun, whereas in the months June and July it is angled towards it. This naturally has an effect on the grey line. See Fig 6.1.

There are a number of ways in which the grey line can be predicted. However, one of the most popular is a simple but effective 'calculator' called the *DX Edge*, and manufactured by Xantec Inc. It is available from a number of outlets including the RSGB in the UK. It consists of a plastic slide rule, one portion containing a map of the world and the other a plot for a given month of the grey line. By setting sunrise or sunset to the given location, it is possible to see the grey line and locations from where signals may travel.

IONOSONDES

The state of the ionosphere has a great bearing on HF radio communications. To assess its state, a piece of equipment known as an *ionosonde* or *vertical incidence sounder* is used. This piece of equipment is very similar to a radar unit which sends out pulses of radio energy and monitors the reflections which are received. The signal is transmitted vertically upwards and the time for the reflected signal to be received enables the height where the

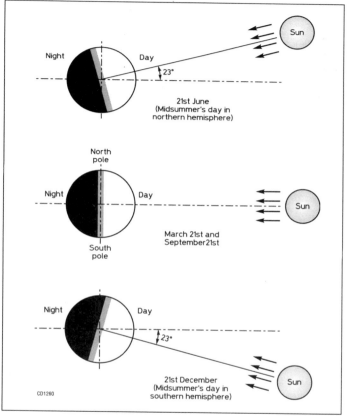

Fig 6.1. The grey line in different months

reflection occurred to be calculated. The frequency is also increased to enable the state of the ionosphere at different frequencies to be assessed. Eventually a point is reached where no signal is reflected back. The output from these ionosondes is generally plotted in the form of an *ionogram* (Fig 6.2).

Naturally these ionograms only indicate the state of the ionosphere above the station. To gain a complete picture, a number of stations are required around the world. Although the number is

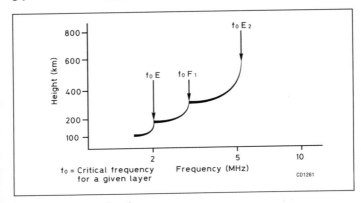

Fig 6.2. Diagram of an ionogram

changing there are currently about 100 fixed stations around the globe, and many other people have access to ionosondes.

PREDICTING CONDITIONS

A number of factors affect propagation conditions, and to gain an accurate view many of them need to be taken into account. However, it is possible to take an approximate view which is sufficient for many listeners and radio amateurs.

To gain a quick view of what conditions may be like there are a number of rules of thumb which can be applied. The position in the solar cycle naturally has a great bearing on conditions, particularly those at the top end of the HF spectrum. During the sunspot maximum frequencies well over 30MHz will be open for ionospheric propagation. Sometimes signals can be heard on frequencies of 50MHz and more via F_2 layer propagation. However, frequencies of between 15 and 20MHz may be propagated at the trough.

On a day-to-day basis it is found that the daily figures of solar flux and geomagnetic activity give a good indication. Solar flux on its own can give a broad indication, and a little experience will link the values of solar flux and band conditions. It can vary from around 65 at the minimum point of the cycle to over 300 at the peak. For good conditions on the higher-frequency bands a flux of around 100 is required.

Table 6.2. Possibility of magnetic storms

Conditions	A Index	Potential of storm
Quiet	0–7	Low
Unsettled	8–15	Low
Active	16–29	Moderate
Minor storm	30–49	Moderate
Major storm	50–99	High
Severe storm	>100	Very high

To gain a fuller indication of conditions the level of geomagnetic activity should be taken into account as well. This indicates the likelihood of a degradation of conditions due to geomagnetic disturbances and storms. A summary of the likelihood of disturbances is given in Table 6.2. As a rough guide it is found that conditions on the high-frequency bands are best when the level of solar flux is high and the magnetic activity level is low.

It is also worth noting that the rotational period of the Sun has an effect on conditions. The solar cycle is 27 days due to the Sun's rotation relative to the Earth, and it is found that conditions tend to repeat themselves with this cycle. If there is a storm, conditions will degrade again after this period. As the sunspots are not evenly spaced around the Sun it is found that solar flux varies. Similarly improvements in conditions from higher values of solar flux also have the same period.

Propagation programs

There are several computer programs which are available to predict propagation conditions. They vary in cost but many can be obtained quite cheaply or downloaded from the Internet. In general they take in the date as well as the solar flux. Some, but not all, also take in information about geomagnetic activity. Using this information they calculate the probability of various paths being open at certain times.

The programs have been generated by a number of organisations. Those like IONSOUND, MINIMUF and MINIPROP were developed as commercial packages but others like VOACAP and PROPMAN were developed by specific organisations. PROPMAN was originally intended for military applications but now with the end of the cold war it has been released onto the commercial market. Whilst many of these programs were written initially for professional use, many also include facilities which are directly intended for radio amateurs, for example giving countries by callsign prefix etc.

Whatever the program which is used, it should be remembered that it only gives a probability of a given path being open as it cannot include all the variables which would need to be taken into account to give a complete forecast.

When obtaining a program it is necessary to choose the right one for your computer's abilities. Some of the older DOS programs give a good account of themselves and are able to run on lower-specification PCs. Many of the newer programs require higher-specification machines but give a very good Windows-based interface.

Specific details of programs are not included here in view of the fact that they are continually being updated and costs and sources are changing.

Chapter 7
OTHER IONOSPHERIC MODES

There are several other ionospheric modes of propagation which have not already been covered. They are not as widely used as the other modes for a variety of reasons – they may only occur infrequently or they may not be easy to predict. As such they are not normally used by professionals, but despite this they are often of great interest to the radio amateur or short-wave listener. Some of these modes provide a form of propagation when normal ionospheric propagation may not be possible, or over a path or at a time when propagation may not be expected. Whilst they may be of great interest to the amateur, the professional user also needs to be aware of them because they may give rise to interference to established links.

In professional circles it is not normally possible to collect sufficient data about modes which occur in a relatively sporadic nature. As a result amateurs have been providing most of the research effort. Often amateurs will be able to make observations of the modes in a way which would be costly for professionals. This is particularly true when there are contests in progress because it is possible to collect large amounts of data about the paths over which contacts have been made.

SPORADIC E

One of the best-known of these occasional propagation modes is *sporadic E* (E_s). In the past television services which used the low part of the VHF spectrum suffered high levels of interference when it was present. Even now the VHF FM band may be affected by it and when it occurs stations from several hundred kilometres away can often be heard.

Sporadic E arises when very intense clouds of ionisation develop around the altitude for the E layer. The levels of ionisation are up to five times the density of the maximum levels achieved in this layer at the sunspot peak. Being very much more intense than normal, these clouds are able to reflect frequencies well into the VHF region.

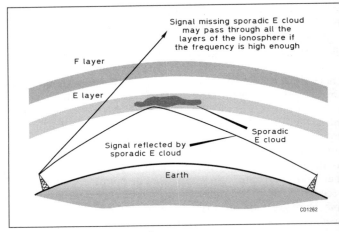

Fig 7.1. Sporadic E propagation

When sporadic E clouds form, the level of ionisation steadily builds up. First it affects frequencies at the lower end of the HF spectrum but, as the level of ionisation increases, so does the maximum frequency which can be reflected. Not all clouds reach the same level of ionisation, so the maximum frequencies which can be reflected vary from one cloud to the next. The 15m and 10m bands are often affected, and the 6m band less so. Sometimes the 2m band (144MHz) can be affected, giving rise to HF like propagation conditions. However, openings as high as this are considerably less frequent and do not last as long. A long opening at this frequency is an hour or possibly a little more, but sometimes openings may only last a few minutes.

The clouds are relatively localised – the horizontal size varies quite considerably from a few tens of metres up to 200km. The shape of the clouds also varies from long and thin to quasi-circular, and they are not very deep, typically being only a few hundred metres thick. They develop in an apparently random fashion, giving rise to the name for this mode of propagation, and are also blown about by the winds in the upper atmosphere, causing the areas which are affected by the propagation to change. This makes this mode of propagation appear even more random. The speed f the clouds may be up to 400km per hour, resulting in skip hanging significantly over a short period of time. This may be particularly noticeable at VHF where higher-gain, and hence narrower-beamwidth, antennas are used. During the course of an opening it may be necessary to move the heading of the antenna a number of times, and it will be noticed that the country from which the signals can be heard or contacted may change. Signal strengths may also change very rapidly. They may be very strong one minute and completely fade out the next.

In view of the fact that the sporadic E clouds are generally slightly below the normal altitude for the E layer the maximum distances which can be achieved are around 2000km. However, double-hop sporadic E openings have been reported on the lower-frequency bands. This is much less likely to happen on 2m because of the significantly lower times which these openings occur.

It is also found that the very high levels of ionisation give very-low-loss reflections, resulting in high signal strengths and enabling stations with low powers or poor antennas to make contacts. However, on 2m the sheer number of high-power stations often reduces the chances for stations with poor antennas and lower powers.

Whilst VHF users often look for sporadic E openings, it can degrade communications on the HF bands at times. Under conditions when normal propagation exists via the F_2 layer considerable distances will be achieved. However, if a sporadic E cloud forms then its intense ionisation will prevent HF signals penetrating. When this happens only short-range signals will be detectable over the area affected by the cloud.

Trying to predict sporadic E is very difficult. However, statistical data has been assembled about its occurrences and this can be used to detect the most likely times for it to occur. It occurs mainly in summer, and in the northern hemisphere the months from May to August yield the highest number of openings, with June yielding the most as shown in Fig 7.2. A small peak is also noticed in December. A similar pattern is also apparent in the equivalent months, November to February, in the southern hemisphere. Generally the frequencies well into the VHF portion of the spectrum are only affected in the middle of the sporadic E season, ie mainly in June and July in the northern hemisphere. It

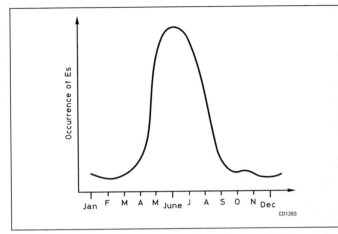

Fig 7.2. Occurrence of sporadic E during the year in the northern hemisphere

should be noted that this only applies to the temperate zones. The occurrence of sporadic E is different in the polar regions and in equatorial regions.

The time of day also has an effect as shown in Fig 7.3. There are two main peaks that occur. One is just before midday, while the other occurs around 7pm. In the afternoon there is a slight fall in the number of openings but at night and during the early morning considerably less have been detected.

Geographical location also has an effect. It is found that more openings are detected in equatorial and polar regions than in the temperate latitudes. This gives rise to the thought that the way in which sporadic E occurs in polar, equatorial and mid-latitudes is different.

The number of openings has also been measured over the duration of the sunspot cycle. From this it might appear that the number of sporadic E openings increases during the periods of the sunspot minima. This would certainly appear to be borne out by the large number of openings noticed in the summer of 1996 at the trough at the end of cycle 22.

The mechanism that causes sporadic E is not fully understood

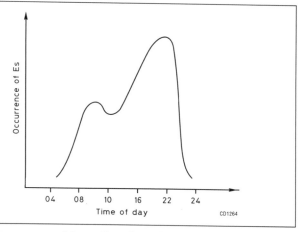

Fig 7.3. Occurrence of Sporadic E throughout the day

yet. A number of theories exist and many scientific papers have been written. Some suggest that it might be related to the level of ultra-violet radiation since it occurs mainly during the hours of sunlight, whilst others indicate there may be a link with electrical storms. The occurrences in the winter at night have also been linked to auroral activity. Other theories suggest that shearing forces caused by the fast moving winds in the upper atmosphere may give rise to these intense clouds of ionisation.

More data is being collected regarding its occurrence and this is likely to increase our understanding of this phenomenon and enable predictions to be made more accurately.

TRANSEQUATORIAL PROPAGATION

Another form of propagation that is used by radio amateurs is called *transequatorial propagation* (TEP). This mode of propagation would have been unlikely to have been noticed except for the observations of radio amateurs. After it was first noticed it was studied almost exclusively by radio amateurs who undertook an investigation which started in 1966. Since then much has been learned about this interesting mode, but there is still much more to find out.

It is found that propagation is supported by the F_2 layer up to frequencies in excess of 100MHz in a north/south direction at times when the normal maximum usable frequency is considerably below this. Typically the MUF for TEP is about one and a half times that for normal modes of F_2 layer propagation.

Openings are particularly noticeable on bands like 50 and even 144MHz where no long-distance propagation may be expected at these times. It has even been recorded at 430MHz. The HF bands are also affected, supporting propagation over a north/south path. This means that there can be significant north/south activity on the HF bands when normal conditions are fading or no longer exist.

Path lengths can range between 2500km and about 5000km but the path must cross the magnetic equator. The two ends of the path should also be approximately equidistant from the magnetic equator. Essentially TEP is a night-time mode with openings normally occurring in the late evening. Typically they occur between 8pm and 11pm with propagation improving until about 10pm after which conditions start to fade. Although it is a north/south (or south/north) mode, angles of up to 20° away from this have been noted on rare occasions.

Like the normal ionospheric modes of propagation TEP is also sensitive to solar activity. It is found that it occurs between 25 and 30 days a month on 50MHz during the months of July to October, disappearing in November. In the years of the solar minimum it peaks in September, but only occurs between 15 and 20 days a month.

It is thought that TEP is caused when an increased level of ionisation occurs in equatorial regions. This enables signals which enter the ionosphere at the correct angle to be propagated across the Equator. In view of the way which signals are reflected they must enter almost in a north/south direction otherwise propagation is not supported. The signals enter the F_2 layer and instead of being reflected to ground in the normal manner they undergo chordal-hop reflection as shown in Fig 7.4.

SPREAD F

Under some conditions the echoes which are received back from ionosonde soundings of the ionosphere indicate that there are

OTHER IONOSPHERIC MODES

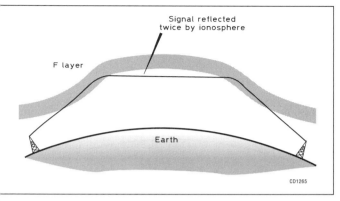

Fig 7.4. Transequatorial propagation

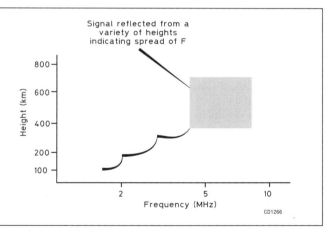

Fig 7.5. Diagram of an ionogram showing spread F

irregularities in the F layer. Instead of receiving a definite echo to give the effective height of the ionosphere at that frequency, a diffused or fuzzy echo is received. This phenomenon is known as *spread F* and is very common in some areas (Fig 7.5).

When spread F occurs the F layer does not consist of a layer with a uniformly varying level of ionisation as normally seen.

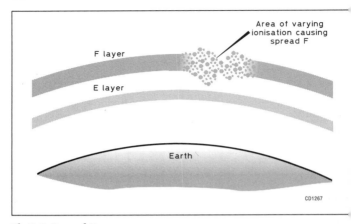

Fig 7.6. Spread F

Instead it appears as though turbulence has broken it up into a variety of areas of differing levels of ionisation, and these each reflect signals, giving a variety of paths which the signal can take (Fig 7.6). Because the signal takes slightly different amounts of time to travel the different paths, the signal at the receiver is a combination of a variety of components which have taken different paths. This results in distortion, and the flutter which is characteristic of signals which have travelled via a polar route. Signal levels are also lower because this method of reflection is not as efficient as normal F layer reflections.

Spread F occurs in two main areas. The first is around the equator between ±20°, and invariably occurs at night. The second is at much higher latitudes above 40°, increasing with latitude to the extent that it is almost permanent in winter. This occurs mainly at night time, but also sometimes in the day. There is little evidence of spread F occurring between latitudes of 20° and 40°.

Interestingly the occurrence of spread F differs between the two regions. In equatorial regions spread F occurs on magnetically quiet days, and disappears with the occurrence of a magnetic storm. However, at higher latitudes it is linked with magnetic activity.

Chapter 8
TROPOSPHERIC PROPAGATION

Although the ionosphere is often thought of as being the main area which affects radio, the troposphere also has a significant effect, athough mainly on frequencies above 30MHz. As such, tropospheric propagation is of far greater interest to the VHF and UHF user.

Most of the effects which are of interest usually take place at altitudes of up to 2km. As these layers are greatly affected by the weather this means that there is a very close link between it and propagation on these frequencies.

LINE OF SIGHT

Under normal or flat conditions communications at VHF and above are often thought to only be along the line of sight. This is not strictly true as signals can be heard over much greater distances than this. There are several effects which come into play at this point – as radio waves and light waves have many of the same characteristics it is possible to see the effects using light which is easier to demonstrate and with which more people are familiar.

The main effect which comes into play is *refraction*. This effect is easily demonstrated with light. It is found that a light ray moving through an area where the refractive index of the medium changes will bend towards the region of the higher refractive index as shown in Fig 8.1. This is easily demonstrated when a stick is placed into water as shown – here the stick appears shorter and to be bent.

The effect can also be seen in air when light rays are bent – it sometimes results in a *mirage*. Tales are told of travellers lost in the desert walking towards water that is not actually there. This is caused by air heated by the hot sand bending the light rays so that a view of the sky is seen. This gives the effect of a distant shimmering pool of water. A similar effect can also be seen above a hot road in summer where the air close to the road surface is heated, causing the shimmering effect to be seen just above the

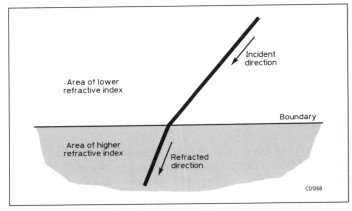

Fig 8.1. Refraction of an electromagnetic wave

surface. This is caused by the light rays bending as the air of differing refractive index rises.

The same effects can be noticed with radio waves, but because water vapour also has an effect on the refractive index for radio signals, the results can be more pronounced.

For most calculations the dielectric constant of air is taken to be unity, but in reality it changes by a very small degree. The average value is about 1.0003, varying under normal conditions between about 1.00027 and 1.00035. The value then falls by about 0.00004 per kilometre. From this it can be seen that the area of highest refractive index is near the Earth. As a result radio waves tend to bend towards the area of higher refractive index and the effect helps the signal to follow the Earth's curvature. As a rough guide it is assumed that the range of a radio signal may be extended by about one third under normal conditions.

ENHANCED EFFECTS

At certain times signals can be heard over distances well beyond what is normally experienced. This often results from increased differences in the refractive index at different points along the signal path. In turn this results in increased levels of refraction.

If the gradient rises above a certain figure then signals are refracted at a rate greater than the curvature of the Earth and they

TROPOSPHERIC PROPAGATION

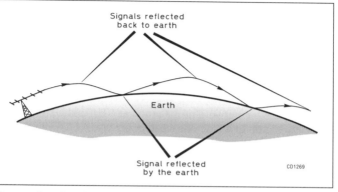

Fig 8.2. Signals refracted back to Earth

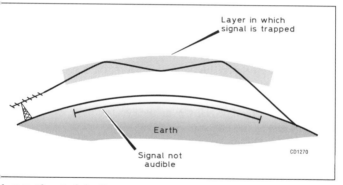

Fig 8.3. Elevated ducting

are returned to the ground where they are reflected upwards again in much the same way that occurs on the HF bands (Fig 8.2). This is one form of *tropospheric ducting*.

Under some conditions signals can become trapped in an elevated duct where there are well-defined changes in refractive index (Fig 8.3). When this occurs signals may travel for several hundred kilometres without being audible on the ground below the transmission path in a similar way to the dead or skip zone experienced on the HF bands using ionospheric propagation.

Extended distances using tropospheric propagation are less pronounced on the lower frequencies in the VHF spectrum. The 144 and 430MHz bands are the most popular as well as many of the higher-frequency bands. Sometimes results can be very spectacular with distances of up to 2000km being achieved.

CORRELATION WITH THE WEATHER

The effects which govern propagation at VHF and UHF occur in the same region in which our weather occurs. As a result there is a strong link between the two and it is possible to monitor the weather maps to gain a view of what propagation conditions will be like.

One of the main ways in which increases in the change in refractive index occur is when a temperature inversion and or a humidity inversion occurs. Under normal conditions the temperature in the troposphere falls with height. However, under certain conditions an inversion can be created which gives rise to a very sharp change in refractive index. Often this occurs when a high pressure area is present and it is particularly dramatic in summer when temperatures are higher and there are correspondingly high levels of humidity. In view of the fact that high-pressure systems are normally stable and tend to move slowly, these conditions may last for many days. However, the greatest improvements tend to occur just as the pressure is falling.

Another common form of temperature inversion occurs with the approach of a cold front. Here the warmer air rises over the colder air beneath, creating a temperature inversion. As these fronts move relatively quickly any lift in conditions will be comparatively short lived.

A number of other conditions can lead to the creation of a temperature inversion. In some conditions the Earth's surface may cool faster after a hot day as the air higher up remains at a more stable temperature. Frosty mornings also result in the air close to the Earth becoming cooler than that higher up. A summer sunrise can give rise to a short-lived inversion, as the air higher in altitude receives the effects of the Sun's rays sooner. Also the subsidence of colder air into a valley can give the same effect.

In view of the close correlation of propagation at VHF and

above it is possible for the experienced user to gain a good idea of what conditions may be like by studying the weather map.

IDENTIFYING THE ONSET OF GOOD CONDITIONS

Whilst it is possible to predict when good conditions are likely to occur, there is no guarantee that a high-pressure area, for example, will bring them about. It is best to look for the likelihood of them occurring on the weather map and then monitor the frequencies of known distant stations. The variety of VHF and UHF beacons which are active give a good indication of an opening on a band. Patterning on UHF television signals is another good indicator as it shows that interference is present. This may be caused by a distant television station being received as a result of the enhanced propagation.

It should also be remembered that not all bands will necessarily be affected. Also conditions may be enhanced in one direction more than another. This can be as a result of the positioning of the enhanced areas of propagation. It is therefore advisable to monitor a variety of beacons in as many directions as possible. On some occasions stations may be heard making long-distance contacts with stations which are not audible. This may be as a result of elevated ducting and the signals may be passing overhead. When this happens continued monitoring may result in the stations being heard as the propagation paths change.

FADING

One of the features which are noticed with tropospheric propagation is that signals are often subject to considerable degrees of fading. Often signals which are heard over considerable distances will be very strong one minute, disappearing into the noise some time later. This occurs because the conditions affecting the propagation are constantly changing. Winds move the air and this will cause the path supporting the signal propagation to change. It is also likely that the overall signal at the receiver will have reached the receiver via a number of different paths. As these change the path length varies and the signals reaching the receiver via the different paths will pass in and out of phase with one another. Sometimes they will combine constructively to enhance the strength of the signal whereas at other times they will interfere

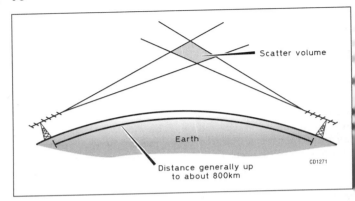

Fig 8.4. Troposcatter

destructively to cancel one another out and reduce the signal strength.

TROPOSCATTER

Tropospheric scatter is sometimes used for communications over distances of up to about 800km. It can be used almost regardless of the conditions, although signal strengths are normally very low. This means that high powers, high antenna gains and sensitive receivers are required. This method of propagation is often used commercially, normally on frequencies above 500MHz. It provides a reliable method of communication, and is much cheaper than using satellites. In fact it provides much of the data communication for the oil rigs in the North Sea when they need to communicate with stations on mainland Britain.

This form of propagation relies on the fact that there are areas of slightly different dielectric constant in the atmosphere at an altitude of between 2 and 5km. Even dust in the atmosphere at these heights adds to the reflection of the signal. A transmitter launches a high-power signal, most of which passes through the atmosphere into outer space. However, a small amount is scattered when is passes through this area of the troposphere, and passes back to the ground at a distant point as shown in Fig 8.4. Path losses are very high, and the angles through which signals can be reflected are normally small.

The area within which the scattering takes place is called the *scatter volume*, and its size is dependent upon the gain of the antennas used at either end. In view of the fact that scattering takes place over a large volume, the received signal will have travelled over a variety of paths. This tends to 'blur' the signal and make high-speed data transmissions difficult.

It is also found that there are large short-term variations in the signal as a result of turbulence and changes in the scatter volume. As a result commercial systems using this type of propagation use *multiple diversity systems*. This is achieved by using vertical and horizontally polarised antennas as well as different scatter volumes (*angle diversity*) and different frequencies (*frequency diversity*). Control of these systems is normally undertaken by computers. In this way these systems can run automatically, giving high degrees of reliability.

ATTENUATION BY THE ATMOSPHERE

For transmissions in the VHF section of the spectrum, atmospheric conditions like rain and fog have little effect on the signals. However, as the frequency increases, the atmosphere has a much greater effect on the level of attenuation in the signal path and at certain frequencies the loss introduced has to be considered.

For most UHF communications the effects of rain and moisture in the atmosphere do not have to be considered as far as any additional path loss is concerned. However, as the frequencies rise above about 3GHz the loss can introduce an additional degree of variation into the path. As may be expected, the loss is dependent upon the amount of rain and also the size of the droplets. As a rough guide, very heavy rain may introduce an additional loss of about 1dB per kilometre at around 5GHz, and more at higher frequencies.

The loss occurs for two reasons. The first is absorption by the rain droplets. The level of actual attenuation is dependent upon the droplet size. The second occurs as a result of the signal being scattered, and although the power is not lost, not all of it travels in the original direction it was intended. In this way the antenna gain is effectively reduced. At frequencies in excess of 100GHz the levels of attenuation are even higher and they are also governed by the amount of water vapour in the atmosphere.

Chapter 9
METEOR SCATTER

Meteor scatter is a form of propagation which is used for professional communications as well as by radio amateurs. Although the ways in which they make use of this interesting form of propagation vary, the basic concept is still the same. In essence the highly ionised trails caused by the meteors as they enter the Earth's atmosphere are used to reflect the radio signals (Fig 9.1). Although the area of the trails is small and means that only a small proportion of the transmitted signal is reflected, there is just enough to be received at a distant point.

METEORS

Every day the Earth's atmosphere is bombarded by many millions of meteors. Most of them are very small – typically they are only the size of a grain of sand. Even ones this size can produce a small visible trail, but one about the size of a small pebble would produce a trail which would be visible for a few seconds. Even larger meteors hit the atmosphere from time to time. Fortunately those which are large enough to penetrate the atmosphere and reach the ground are few and far between. Meteors fall into two

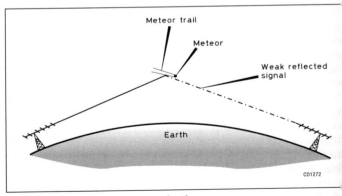

Fig 9.1. Meteor scatter communication

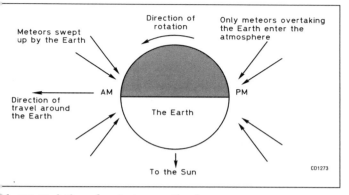

Fig 9.2. Variation of meteors over the day

main categories – those from meteor showers and those which occur sporadically.

SPORADIC METEORS

Meteors enter the atmosphere at all times of the year. They arise from space debris in the universe. Most of the sporadic meteors are thought to come from the Sun which throws out huge amounts of material.

The rate at which meteors enter the atmosphere is far from constant even after the meteor showers are discounted. This arises for a number of reasons. It is found that the part of the atmosphere rotating into the sunrise receives more meteors than that rotating into the sunset as shown in Fig 9.2. The reason for this is analogous to a car travelling in a rainstorm where the front windscreen receives more rain than the rear one. The meteors which enter around sunset have to be travelling faster than the Earth to be able to catch up. Those which enter around sunrise are swept up as the Earth moves around the Sun.

This effect means that the maximum number of meteors enter the atmosphere around 6am and the minimum around 6pm. The ratio of the maximum to minimum is dependent upon the latitude but is around 3:1.

A similar but smaller effect arises with the changing seasons. As the seasons change the angle of tilt of the Earth changes and

Table 9.1. Major meteor showers

Shower	Begins	Maximum	Ends
Quadrantids	1 January	3 January	6 January
April Lyrids	19 April	21 April	24 April
Eta Aquarids	1 May	4 May	7 May
June Lyrids	10 June	15 June	21 June
Ophiuchids	17 June	20 June	26 June
Capricornids	10 July	26 July	15 August
Delta Aquarids	15 July	27 July	15 August
Pisces Australids	15 July	30 July	20 August
Alpha Capricornids	15 July	2 August	25 August
Iota Aquarids	15 July	6 August	25 August
Perseids	25 July	12 August	18 August
Orionids	16 October	21 October	26 October
Taurids	20 October	4 November	25 November
Cepheids	7 November	9 November	11 November
Leonids	15 November	17 November	19 November
Geminids	7 December	14 December	15 December
Ursids	17 December	22 December	24 December

different areas of it present a larger frontal area to the oncoming meteors. As a result the southern hemisphere receives its maximum number of sporadic meteors around March and the northern hemisphere around September.

A third variation has been seen to fall in line with the 11-year sunspot cycle. Evidence suggests that the number of sporadic meteors reaches a peak at the minimum of the cycle, resulting in meteor scatter propagation reaching its peak efficiency when the HF bands are at their minimum, and vice versa.

METEOR SHOWERS

To radio amateurs meteor showers are the best time for meteor scatter propagation. At certain times of the year the number of meteors entering the atmosphere rises quite considerably.

Table 9.1 lists the major showers. Some of these are relatively small, but the Perseids in August is renowned as producing a large number of meteor trails, giving a good visual show as well as providing a large number of trails for radio communications.

Meteor showers are thought to be the result of comet trails. Most of the major meteor showers have been linked to specific comets. These showers are collections of particles which orbit the Sun in an elliptical orbit, and as a result the Earth passes through the collection of particles at the same time each year.

When looking at meteors from a shower it will be seen that they appear to come from a particular point in the sky. This is only a perspective effect and is caused by the fact that the meteors enter the Earth's atmosphere parallel to one another. However, this enables the showers to be named. The point where the meteors seem to appear is called the *radiant*, and this point can be used for naming the shower. For example, the Perseids shower has its radiant in the constellation of Perseus.

METEOR TRAILS

Meteor trails form when meteors of all sizes enter the atmosphere. The meteors enter at speeds varying anywhere between 10 and 75km per second, burning up at altitudes of between 80 and 120km. The height at which the meteors burn up is dependent upon their size, weight and the angle at which they enter the atmosphere. For a given size of meteor, those travelling with a higher velocity will tend to burn up at higher altitudes, as more heat is generated. Those with lower velocities will only generate sufficient heat as the air density increases. Many trails at all altitudes are visible on a clear night and it is often possible to see a good number especially after midnight when there are more.

The trails are formed as the friction of the air causes the meteor to heat up to such a degree that the atoms vapourise to leave a trail of positive ions and negative electrons. The trail is a long thin parabola with the meteor at its head, typically between 15 to 40km long but less than 5m wide.

Once the trails are generated they come under the influence of the winds which exist at these altitudes. When they are formed they are straight, but horizontal winds of around 25m per second and vertical shearing currents of up to 100m per second quickly distort them. It is not uncommon for a section of a trail to be rotated through an angle of around 5° in a second.

The trail has a very high density of free electrons. It is much greater than that generated by solar radiation in the ionosphere, and as such it is possible for the trails to reflect radio signals up to frequencies of around 150MHz. However, in view of the very small area presented by the trail, only small levels of power are reflected.

Meteor trails can be categorised into two types: *over-dense* and *under-dense*. Although the changeover point is taken to be an electron density of 1×10^{14} electrons per cubic metre, it is really the way in which the trails react which is the governing factor. Over-dense trails have a higher electron density which prevents complete penetration of the signal, which is reflected using the same principles that cause signals to be reflected in the ionosphere. Under-dense ones have a lower electron density and allow the incident signal to penetrate into the trail which acts as a form of scatterer.

Over-dense trails are not normally used for commercial systems but they are used for amateur radio. They reflect signals for longer than the under-dense ones and they are much less frequent as they require larger meteors. Hence the requirement for radio amateurs to look out for meteor showers. However, the over-dense trails suffer the disadvantage that signals undergo multipath effects which can make them difficult to copy and they produce very large signal variations. Despite this the trails may last for several seconds, allowing significant amounts of data to be transmitted using a single trail.

Under-dense trails are produced by smaller meteors. Signals from these trails usually rise to a peak in a few hundred milliseconds and then gradually fade away over a period from a second or so down to a few tenths of a second. The signals fade as the electrons from the trail spread out, decreasing the density of electrons to a point where they have little effect on the signals.

FREQUENCIES

The maximum frequency limit for meteor scatter is generally around 150MHz although some dense trails will support communication on higher frequencies; up to 500MHz has been known on some occasions. Commercial systems tend to use frequencies at the low end of the VHF spectrum. Part of the reason is that they use under-dense meteor trails that have lower levels of ionisation and cannot support communication on such high frequencies. The other factor is now mainly historical and results from the fact that these systems had to avoid interference to and from the old VHF television transmissions. As such most commercial systems operate in the range 30 to 50MHz. Here the lower limit is set more by the fact that meteor burst communications must generally be above the maximum usable frequency for ionospheric propagation.

Ideally the use of lower frequencies gives better reflections. The use of lower frequencies means that the level of ionisation is able to support communication for longer. There is around a factor of 10 difference between frequencies of 50 and 150MHz. This makes communication on the higher frequencies much more difficult.

DISTANCES AND TECHNIQUES

It is possible to use meteor scatter techniques to communicate over distances from around 400km up to to about 2000km. Below these distances high angles of elevation are required and the system becomes less efficient. Generally the optimum distance is around 1000km.

When making contact it is found that the best beam heading is not along the direct path. Instead the best heading is around 10 to 15° off to one side as this is where the most usable reflections can be achieved.

Antenna gain is an important aspect of the system. A highly directive antenna will improve the system gain and increase signal strengths. However, a high-gain antenna also has a much narrower beamwidth, and the amount of the sky which is illuminated is much less. In turn this reduces the number of ionisation trails which can be used (Fig 9.3). Therefore the antenna gain is a compromise between signal strength and the number of trails which are required.

It is also found that the signals which are propagated by meteor scatter undergo a Doppler shift. This results from the fact that the point of reflection changes as the meteor moves forwards, and the trails behind it diffuses and reflection decays. This can give a shift in frequency of as much as 2kHz. Also signals at higher frequencies are only present for a short time, giving an effect rather similar to a 'ping'. Signals at lower frequencies are generally present for longer.

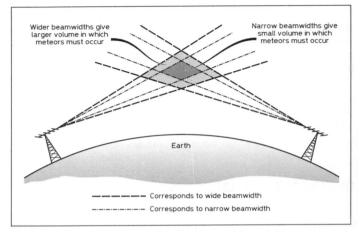

Fig 9.3. Effect of antenna gain on meteor scatter systems

Amateur techniques

Most amateur activity takes place on the 144MHz band, although it is easier to make contacts on lower-frequency bands like 50MHz. Sometimes it is possible to use 432MHz although this is right at the limit for this type of propagation.

Contacts can be made on 50MHz with relatively low power because the reflections are good at this frequency. At 144MHz where higher levels of ionisation are required reflections are much shorter and higher power levels are needed. Typically powers of 100W at the antenna and an antenna gain of just over 10dB is needed. Coupled to this a low-noise front-end is essential because of the low signal strengths at the receiver.

In view of the intermittent nature of the signal paths, specialised procedures are required. Single sideband can be used occasionally on 50MHz, but generally very-high-speed Morse code is used.

Typically speeds of up to 400 words per minute are employed with computers used to code and decode the signals. Previously tape recorders were used to speed up or slow down the signals so that they could be read.

Commercial systems

Most commercial meteor scatter systems are used for non-real-time data communications. Two stations are set up, one sending out a probe signal, and the other monitoring the frequency. When a suitable trail is available, the receiving station picks up the probe signal and sends a reply. Data is then transferred in packets, the receiving station sending out acknowledgements after each packet. When the trail fades and communication is lost the probe signal is sent out again until another usable trial is detected.

In view of the intermittent nature of the paths, meteor scatter is only used for data communications which are not required for a real-time application. For a commercial data system it may take 10 to 15 minutes to send data as this will mean using several different meteor trails. Obviously if large amounts of data have to be sent then longer delays will be experienced.

Now that computer technology is cheap and freely available meteor scatter provides a reliable means of communication which is not subject to the variations of the ionosphere which can totally disrupt communications especially when there is an ionospheric storm.

Appendix
Further Reading and Sources of Data

Radio Communication Handbook, edited by Dick Biddulph, G8DPS, RSGB.

Amateur Radio Operating Manual, edited by Ray Eckersley, G4FTJ, RSGB.

New Short Wave Propagation Handbook, Jacobs, Cohen and Rose, CQ Communications Inc.

Principles of Radio Communication, edited by F F Mazda, Focal Press (imprint of Butterworth Heinemann).

Radio Auroras, Charlie Newton, G2FKZ, RSGB.

There is a table printed each month in the RSGB magazine *RadCom* showing path predictions from the UK to a number of locations around the globe. They are shown in the form of numbers from 0 to 9, indicating probabilities between 0 to 90% of a particular path being open.

The GB2RS weekly news broadcasts give a large amount of propagation indices and data for assessing propagation conditions. These broadcasts are made on a variety of frequencies on a Sunday. Details of frequencies and times are available in the latest *RSGB Yearbook*.

For anyone with access to the Internet there is a vast amount of information about propagation that is available. Data is frequently updated and there is an interesting selection of articles. A few site addresses are given below:

http://www.keele.ac.uk/depts/por/psc.htm
http://www.sec.noaa.gov
http://www.rsgb.org
http://www.ips.gov.au/papers/

INDEX

A
A index, 26
Amateur bands, 8, 47
Angle diversity, 71
Appleton, 3
Auroras, 44

B
Bands, amateur 8, 47
Beacons, 51

C
Chapman, 3
Critical frequency, 33

D
D layer, 18
  attenuation, 28
de Forest, 2
Dead zone, 33
Dellinger effect, 43
Diversity reception, 71

E
E layer, 3, 19
  reflection, 29
Edison Petit, 4
Electromagnetic field, 6
Electromagnetic spectrum, 8

F
F layer, 19
  reflection, 29
Fading, 36, 69
Field
  electromagnetic, 6
  strength, 10
Flares, solar 41
Flux, solar, 25
Free-space propagation, 13
Frequency, 6
  critical, 33
  optimum, 35
Frequency diversity, 71

G
Galileo, 4
Geomagnetic storm, 43
Gigahertz, 6
Grey line, 37
Ground-wave propagation, 13, 27

H
Heaviside, 2
Hertz, 6
Hulbert, 3

I
Interference, multipath, 36
Ion, 16
Ionisation, 17, 20
Ionogram, 53
Ionosonde, 52
Ionosphere, 4, 16
  propagation, 13
  storm, 43

K
K index, 26
Kennelly, 2
Kennelly-Heaviside layer, 2
Kilohertz, 6

L
Loss, path, 11
LUF (lowest usable frequency), 33

M
Magnetosphere, 40
Marconi, 2
Megahertz, 6
Mesosphere, 14
Meteors
  scatter, 72
  showers, 74
  trails, 75
Mirage, 65
MUF (maximum usable frequency), 34

Multipath interference, 36
Multiple diversity systems, 71

O

Optimum frequencies, 35
Over-dense meteor trails, 76

P

Path loss, 11
Plages, 23
Polar cap absorption (PCA), 43
Polarisation, 9
Predicting propagation, 47, 54, 69
   computer programs, 55

R

Radiant, 75
Refraction, 65

S

Scatter volume, 71
Schwabe, 4
Short-wave fades, 42
SIDs (sudden ionospheric
      disturbances), 41
Signal strength, 10
Skip
   distance, 32
   zone, 33
Solar flares, 41
Solar flux, 25
Solar wind, 20, 39
Spectrum, electromagnetic, 8
Sporadic E, 57
Sporadic meteors, 73

Spread F, 62
Storms, geomagnetic and
      ionospheric, 43
Stratosphere, 14
Sunspots, 4, 22
   cycle, 4, 24
   number, 4, 23
Sunspot Index Data Centre, 24

T

Taylor, 3
TEP (transequatorial
      propagation), 5, 61
Thermosphere, 14
Transequatorial propagation
      (TEP), 5, 61
Tropopause, 15
Troposcatter, 70
Troposphere, 14
   ducting, 67
   propagation, 13
   scatter, 70
Twilight zone, 37

U

Under-dense meteor trails, 76

V

Vertical incidence sounder, 52

W

Wavelength, 6
Wolf number, 23
Wolf, Rudolf, 23

Your First Packet Station

by Steve Jelly

This handy guide will show you just how easy it is to set up a basic packet radio station and enter the world of data communications from your shack. It describes the equipment you need, and how to connect it up and communicate with other packet users. It also explains the other features of the network such as bulletin boards and DX clusters. Explanations are kept as simple and non-technical as possible, making this book an ideal choice for beginners.

75 pages; 177 by 111mm; 1996

Radio Society of Great Britain, Lambda House, Cranborne Road, Potters Bar, Herts EN6 3JE

Tel: 01707 659015

Visit our website at www.rsgb.org